"Destined to become a classic in the field, this skillfully
fluency and mastery of acceptance and commitment th
amazing. Get it and learn from one of the best."

—**Ole Taggaard Nielsen, ClinPsyD**, Association for Contextual
Behavioral Science (ACBS) peer-reviewed ACT trainer at ACT
Klinikken in Denmark

"In this intriguing book, Robyn shares wisdom gained from many years of clinical practice and teaching. The result is profound guidance for clinicians, researchers, and trainers exploring the interwoven threads of fidelity, competence, and mastery in the practice of ACT."

—**Patti Robinson, PhD**, and **Kirk Strosahl, PhD**, cofounders of
focused ACT, and coauthors of *Real Behavior Change in Primary Care*
and *Brief Interventions for Radical Change*

"*The Heart of ACT* is exactly that—a guide to the essential therapeutic process at the core of ACT. While other volumes focus on theory and technique, Walser illuminates the moment-to-moment opportunities, choices, and experiences that drive growth and change. Through her clinical scenarios, supervision dialogues, and incisive analysis of the treatment process, we are shown, in dazzling clarity, how a master clinician practices ACT."

—**Matthew McKay, PhD**, coauthor of *Acceptance and Commitment
Therapy for Interpersonal Problems*

"The heart and wisdom of ACT, and the importance of creating meaning in every moment, is embodied within every chapter. Filled with insight, depth, grace, and clear steps for bridging the gap between conceptual understanding of ACT and its flexible implementation, Robyn Walser writes with compassionate urgency and expertise gleaned not only from decades of work with clients, but also from her own personal journey. Readers are routinely invited to consider their own experiences and to engage in personal reflections. This rich therapeutic volume is a must-read for anyone who wants to grow as a clinician."

—**Robert J. Kohlenberg, PhD, ABPP**, and **Mavis Tsai, PhD**, cocreators of functional analytic psychotherapy (FAP)

"*The Heart of ACT* is the book I have been waiting for: a book for professionals that speaks to the relationship between the client and therapist, written by a therapist that works with clients regularly. Robyn Walser is a master therapist, and this book allows the reader to step inside her work and learn. It not only helps the reader understand their clients better, it also helps the reader understand themselves and their own behavior in the therapy room. I highly recommend it for anyone wanting to learn to do therapy 'from the feet up.'"

—**Louise Hayes, PhD**, clinical psychologist and senior fellow, University of Melbourne and Orygen Centre for Excellence in Youth Mental Health; and coauthor of *The Thriving Adolescent* and *Get Out of Your Mind and Into Your Life for Teens*

"If you have been fortunate enough to be one of the many thousands of attendees at Robyn's workshops, you almost certainly walked away inspired and in awe of her ability to rapidly get to the heart of the matter in therapy. In this book, Robyn shares herself and her wisdom. It feels like she is personally there guiding you on your ACT journey, grounding you in process, helping you connect with your heart, and inspiring you to mastery. If you are looking for a book to take your ACT work to the next level, this is the book."

—**Jenna LeJeune, PhD**, and **Jason Luoma, PhD**, cofounders of Portland Psychotherapy; peer-reviewed ACT trainers; and coauthors of *Values in Therapy*; Luoma is also coauthor of *Learning ACT*

"*The Heart of ACT* is a crucial book for any and every ACT therapist. Robyn Walser goes beyond looking at ACT in terms of simple therapy tools to find the heartfelt, compassionate, and deep core of ACT processes. This book breathes life into the ACT therapy relationship, straight from the heart."

—**Dennis Tirch, PhD**, and **Laura Silberstein-Tirch, PsyD**, Dennis Tirch is founder of The Center for Compassion Focused Therapy, associate clinical professor at Mount Sinai, and author of *The Compassionate-Mind Guide to Overcoming Anxiety*; Laura Silberstein-Tirch is a clinical psychologist, director of The Center for Compassion Focused Therapy, adjunct assistant professor at Albert Einstein School of Medicine, and author of *How to Be Nice to Yourself*

"I can think of no better person to examine the heart of ACT than Robyn Walser. I've had the privilege of calling Robyn a close friend and colleague for over twenty years, and have long appreciated her formidable talents as an ACT therapist, supervisor, consultant, and trainer. She is all heart—bringing a rare combination of clinical acumen, compassion, and unflinching honesty to her work. This generous book is no exception. It is accessible, on point, and extremely well written. Its unique format offers a rare window into how a master clinician approaches the therapy, and guides others through the many challenges and nuances that arise. I highly recommend it as a must-read for anyone interested in developing and refining their ACT skills—and their heart."

—**Darrah Westrup, PhD**, is in private practice in Durango, CO; and author of *Advanced Acceptance and Commitment Therapy*

"Up your ACT game with the Therapy Whisperer herself. Robyn Walser's words of wisdom will guide you on your own journey of experiential discovery as an ACT provider, a fellow human, and ultimately, a healer. In these pages, clinicians can finally begin to transcend the gap from conceptual understanding to experientially fluid application of ACT processes—from the inside out. As only Robyn Walser can, she will help you find therapeutic presence, with heart and wisdom."

—**Lara E. Fielding, PsyD, EdM**, author of *Mastering Adulthood*, clinical psychologist, and adjunct professor at Pepperdine Graduate School of Education and Psychology

"At last, a book that articulates all the richness that exists between the lines of other ACT manuals and tutorials. Robyn Walser exemplifies the seamless marriage of being and doing, and *The Heart of ACT* illuminates her process, complete with experiential learning activities for deepening your own clinical expertise. This is the volume that has been missing from the ACT canon."

—**Jennifer L. Villatte, PhD**, assistant professor in the department of psychiatry and behavioral sciences at the University of Washington School of Medicine, and coauthor of *Mastering the Clinical Conversation*

"*The Heart of ACT* is one of the first clinical manuals to embrace the new process-based era of CBT, in which psychological issues are no longer approached through an exclusive behavioral, cognitive, or emotional lens—but holistically, as functional contextual processes. Infused with the exceptional human qualities and clinical expertise of Robyn Walser, this book will teach you how to expand your therapist repertoire beyond ACT exercises and protocols to integrate more dynamic and fluid therapeutic interventions."

—**Matthieu Villatte, PhD**, Bastyr University, Seattle; coauthor of *Mastering the Clinical Conversation*

The Heart of ACT

Developing a Flexible, Process-Based & Client-Centered Practice Using Acceptance & Commitment Therapy

---

ROBYN D. WALSER, PhD

With Manuela O'Connell, PhD,
& Carlton Coulter, DClinPsy

Context Press
An Imprint of New Harbinger Publications, Inc.

Publisher's Note

This publication is designed to provide accurate and authoritative information in regard to the subject matter covered. It is sold with the understanding that the publisher is not engaged in rendering psychological, financial, legal, or other professional services. If expert assistance or counseling is needed, the services of a competent professional should be sought.

Distributed in Canada by Raincoast Books

Copyright © 2019 by Robyn D. Walser
                Context Press
                An imprint of New Harbinger Publications, Inc.
                5674 Shattuck Avenue
                Oakland, CA 94609
                www.newharbinger.com

Cover design by Amy Shoup; Acquired by Ryan Buresh;
Edited by Rona Bernstein; Indexed by James Minkin

All Rights Reserved

Library of Congress Cataloging-in-Publication Data

Names: Walser, Robyn D., author.
Title: The heart of ACT : developing a flexible, process-based, and client-centered practice using acceptance and commitment therapy / Robyn D. Walser ; foreword by Steven C. Hayes.
Other titles: Heart of acceptance and commitment therapy
Description: Oakland, CA : New Harbinger Publications, Inc., [2019] | Includes bibliographical references and index. | Identifiers: LCCN 2019017392 (print) | LCCN 2019020622 (ebook) | ISBN 9781684030408 (PDF e-book) | ISBN 9781684030415 (ePub) | ISBN 9781684030392 (paperback)
Subjects: LCSH: Acceptance and commitment therapy. | BISAC: PSYCHOLOGY / Clinical Psychology. | PSYCHOLOGY / Psychotherapy / Counseling.
Classification: LCC RC489.A32 (ebook) | LCC RC489.A32 W35 2019 (print) | DDC 616.89/1425--dc23
LC record available at https://lccn.loc.gov/2019017392

Printed in the United States of America

21   20   19

10 9 8 7 6 5 4 3 2 1   First Printing

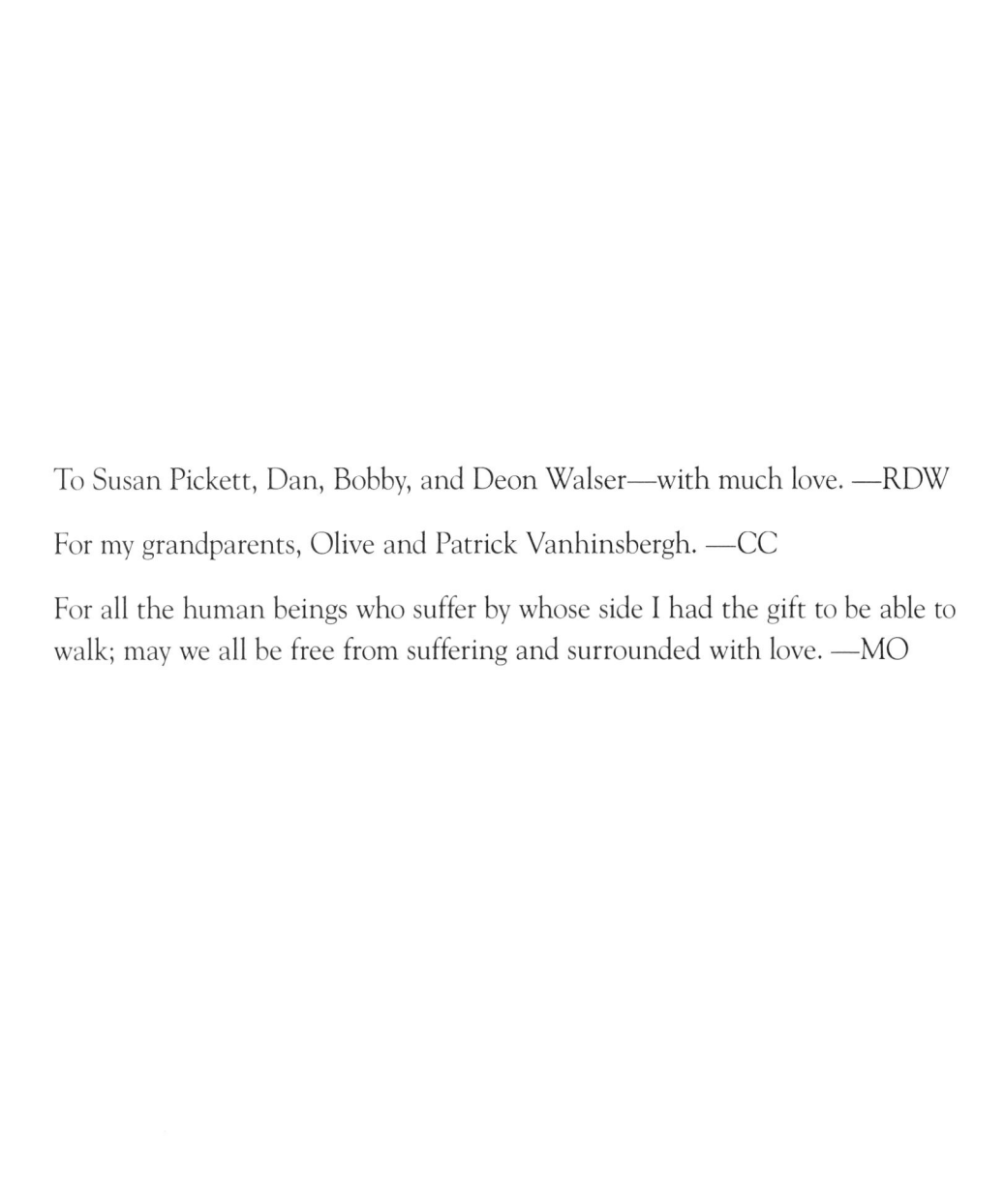

To Susan Pickett, Dan, Bobby, and Deon Walser—with much love. —RDW

For my grandparents, Olive and Patrick Vanhinsbergh. —CC

For all the human beings who suffer by whose side I had the gift to be able to walk; may we all be free from suffering and surrounded with love. —MO

Contents

Acknowledgments		vii
Foreword		ix
Introduction: The Heart of ACT		1

Part 1: Building Heart Through the ACT Core Processes

Chapter 1	Getting *Your* ACT Together	13
Chapter 2	Living Life from the Feet Up: Open, Aware, Engaged	33
Chapter 3	Open and Aware: Knowledge of Self in Relation to Other	55
Chapter 4	Engaged: Existence and Purpose	75
Chapter 5	Growing Your Therapeutic Fluency: What You Say and How You Say It	87

Part 2: Building Heart Through Unfolding Experience

Chapter 6	Engagement in Process	115
Chapter 7	Overarching, Interpersonal, and Intrapersonal Process in Practice	135
Chapter 8	The Clinician's Experience: A Closer Look at the Intrapersonal Process	149
Chapter 9	The Client and Clinician Experience: Exploring Interpersonal Process Through Challenge	171
Chapter 10	The Stance of the ACT Therapist	195

Epilogue: Owning Your Life; Living with Heart	215
References	221
Index	227

Acknowledgments

I would like to thank Steven C. Hayes, my mentor and friend, for his guidance and support over the years. As always, a thank-you to my mom—my hero. And…I would like to offer appreciation to Ruby and Sydney, my dogs, for patiently lying by my side while I wrote this book. A special thanks to Manuela O'Connell and Carlton Coulter for supporting and challenging me throughout the years, for giving their time and wonderful contributions to this book, and for their friendship. —RDW

I would like to thank Robyn for involving me in the writing of this book, Oenone Dudley for her ongoing support for all aspects of my work, and Sophie for creating the time for me to contribute to this project. —CC

I would like to thank my parents for inspiring and teaching me to commit to make this a better world. To my very dear family for all the support and patience in walking by my side in good and rough times. I also would like to thank my students and clients who, as fellow human beings, helped me develop and grow. And finally, a deep gratefulness, from my heart, to Robyn for her loving and inspiring presence all these years. It has had a profound impact on me and my life beyond words. —MO

FOREWORD

Digging Down to the Essence

ACT skills have long been talked about in terms of the therapist's head (theoretical knowledge of the work), hands (technical skill in doing the work), and heart (experiential contact with the work). I have always told my students (among whom I proudly count Robyn Walser) that the last is most important.

I think most expert ACT therapists would agree: if you are going to support others in thinking freely, feeling fully, and focusing on what matters, it is your own experience that will be your best guide. To do ACT well, you need to be fully present with another human being, walking a values-based journey with them. This stretches us as human beings, but it also puts vitality into our work. The heart of ACT helps us fill the gap between topographical fidelity to the model and actual mastery of the model.

There are scores of books on the head and hands of ACT. Despite its importance there are very few books on the heart of ACT.

One reason for this is that books are linear and literal, while experience is complex, recursive, and to some degree *beyond* words. It is an art to use words in a way that evokes transformational experiences. Poets and novelists know how to do so, but if you are trying to get to the heart of an evidence-based therapy, poems and stories alone simply will not do. You need to come at the experiential core of the work repeatedly—from different angles, with different issues, and even with different voices—to orient the reader toward features of their reactions that they can learn from and use.

This book does exactly that.

As you walk through each of these issues and angles, you come a little closer to the heart of ACT–to the actual experiences on which mastery of the model depends. Using personal tales and gentle reorientation to what is important, Robyn asks the reader to notice what is hidden inside our clinical moments and practical clinical struggles. As she does this skillfully in area after area, mere words begin to produce an experiential sense that goes beyond words.

If you look up the synonyms of *heart*, you will see experiential words that express feeling, connection, and action. Words like *character*, *love*, *benevolence*, *compassion*, *understanding*, *sensitivity*, and *tenderness* show up; and words like *boldness*, *bravery*, *guts, purpose, nerve,* and *fortitude*. This is not a bad summary for all of ACT: openness, connection, and bold moves.

I can hear Robyn as I read this book. Even as a student, Robyn was a therapist with amazing heart, boldness, and skill. But in the decades since she received her PhD, Robyn has probably done more clinical trainings and listened to more tapes of ACT sessions than anyone else alive. She helped refine and extend the scoring system for ACT clinical tapes and supervised their use as part of the rollout of ACT into US Veterans Affairs hospitals. She has personally watched countless hours of ACT sessions and supervised therapists across the globe. She has almost a sixth sense about what is going on inside ACT therapists as they attempt to apply the ACT model. She not only embodies the heart of ACT, she is one of the world's experts on noticing its features and feeling its pulse.

As supervisees and experienced therapists, Manuela and Carlton provide counterpoints and amplifications that help the reader see through differences of style to the essence of the work.

Getting to the heart of any subject means getting to its essence. ACT has an essence, and in my view, it is why people are drawn to it. I tell my workshop attendees that if after a day or two of training in ACT you do not sense a connection, it is better to focus on other evidence-based methods. Conversely, I have watched how people develop as ACT therapists for nearly 40 years, and I cannot think of a single person who was deeply connected to the heart of the work, who with persistence and care did not eventually learn the concepts and methods enough to be a very good ACT therapist.

If you are drawn to ACT, you are likely drawn to its heart. That's great, and you have the right book in front of you.

—Steven C. Hayes PhD
    Foundation Professor of Psychology
    University of Nevada, Reno
    Originator and codeveloper of Acceptance and Commitment Therapy

INTRODUCTION

The Heart of ACT

Occasionally in life there are those moments of unutterable fulfillment which cannot be completely explained by those symbols called words. Their meanings can only be articulated by the inaudible language of the heart.

—Martin Luther King, Jr.

Acceptance and commitment therapy (ACT; Hayes, Strosahl, & Wilson, 2012) has transformed the lives of clients and clinicians in many ways and—as part of a broader shift in the practice of psychotherapy emerging from a new theory of human language (see Hayes, 2004)—has had a significant impact in the field of psychology. ACT has flourished in areas of application, research base, and clinical use around the world. Not only does it have a broad reach that continues to grow, it also tends to have a distinctive and personal impact. Indeed, ACT has had a significant influence in my personal life and has been my therapeutic intervention of choice. It is a passion that I wish to share with others who are interested in learning the approach. Helping others to digest the theory and research behind ACT and understand its content, processes, techniques, and foundational goal—psychological flexibility in the service of creating a meaningful life—has been hugely rewarding, both in training experiences and in the clinical setting. Additionally, connecting others to ACT from a more in-depth, experiential, or heartfelt place—born out of the knowledge of human joy and suffering and the movement toward the inevitable end of our personal existence—has also been a welcome part of my journey.

From its inspiration to its fruition, this book has been about exploring the gift that ACT can bring to clinician and client alike. The writing herein, from both a professional and personal standpoint, reaches beyond a simple notion of a psychological approach. ACT, in its fullest form, stretches beyond the words we use to describe it. Done with intention and presence, ACT links us to the very qualities of what it means to be alive and whole, to be a conscious and experiencing being. And with this, one can recognize a significant sense of "heart" in the work done in ACT. That is, connecting to a broader sense of self and openly observing the movement of experience in each moment allows the therapist not only to recognize their own wholeness, their own capacity to be open and active, but also to recognize this in others. The therapist is a whole and experiencing being, and the client is as well. This recognition frees the therapist to interact with the client from a different perspective, one in which no matter what emotion, thought, or sensation is experienced (by client or therapist), it does not diminish the other as a whole being. It is here that you and your client are acceptable, loved, and free. This is the true heart of ACT.

The most challenging part about this description of the "heart" of ACT is that it is, in part, a felt experience; it cannot be adequately conveyed in the promise of a book. It is both consciousness and unconditional presence and is, therefore, free from impatience, judgment, and all things selfish and "mind-y." It is not meant to be mystical; however, it may be transcendent or beyond merely knowing with the mind. Contacting this place, then, will mean cultivating personal awareness. But the invitation, here, is to go deeper into this practice, touching the unaffected stillness that is "you."

Wisdom in ACT is also essential. Building this wise self involves meditative and self-reflective practices, truly seeing yourself as you are, and wholly recognizing the inevitable rise and fall of all life, including your own. This ability to combine a thoughtful manner with reflection and compassion can be cultivated, improving your personal well-being across time (Ardelt, 2004). Wisdom also involves flexibility in behavior, being curious and open in the process of making choices in life. More specific to work with clients, I am referring to the therapist's capacity to discern (through thoughtfulness and reflection) and then deliver, in an authentic fashion (from a place of compassion), the intervention that supports behavior change in the service of personal values. Wisdom stands in recognition that consequences will be present no matter what choice an individual makes. In ACT, we hope that these choices will ultimately be linked to purpose. And purpose, I would argue, is only found in contacting the knowledge of our own death. Sitting with awareness of choice, consequences, and death may require being bold in life and therapy. Sometimes it is difficult to be authentic and still empathic. It can be challenging to deliver needed consequences and say the thing that is hard to say. This may require you to be aware of your own and others' tendency to fuse and avoid, and to act on it *as needed* within the context of a therapeutic relationship to promote change.

Heart and wisdom, together then, make a kind of therapeutic presence in ACT. It is this presence, in combination with a high degree of facility in implementing the six core processes (Luoma, Hayes, & Walser, 2017) as processes, not simply as techniques, that supports competence and mastery in ACT. For some therapists, though, shifting to this kind of presence and delivering ACT as a multilevel, highly experiential process has proved difficult. It's understandable why. Training therapists in ACT is a challenging task because it involves integrating (1) a complex combination of therapeutic, observational, and behavior analytic skills; (2) ACT processes; and (3) broader intrapersonal, interpersonal, and change processes such that they are readily organized, accessed, and implemented in different contexts while staying inside of the heart and wisdom.

Additionally, new challenges to this endeavor have arisen. Dissemination, along with its inevitable dilution of the intervention being taught, has pushed the techniques and exercises conducted in ACT to the forefront, leaving the process and processes of ACT behind. This book is an attempt to turn the tide. It's possible to see how the parts (separate core processes) fit into the whole (psychological flexibility) and how they fluidly link one to the other in the flow of a process-based therapy. But reconnecting to ACT's heart, to ACT's wisdom, means reorienting to what it means to be in therapy as one human being interacting with another in the ongoing flow of purposeful change. This book is geared to the therapist and intended to bring a gentle, compassionate, and hopefully fun learning experience to the reader in the service of this reorientation. I hope you'll find the explorations of therapeutic presence and process in therapy useful in your competent implementation and movement toward mastery of ACT—discovering, in the end, where this heartfelt presence lies.

ACT and Its Training Challenges

In leading ACT training and providing supervision for more than twenty-five years, and in listening to hundreds of taped sessions, I have seen and heard both the beautiful delivery of ACT and the awkward fumbling of first attempts at metaphors and exercises, demonstrating an expected range of skill in delivering the model. However, as the use of ACT has continued to spread, I have come to notice a gap in the translation from theory to practice, from workshop training to clinical settings, and from ACT fidelity to competency and mastery. Dr. Darrah Westrup's book, *Advanced ACT* (2014), is a must read for clinician's working to close this gap. Her clinical acumen and thoughtful writing about challenges in learning ACT and assisting clinicians to focus on process is the work of an advanced ACT clinician. *The Heart of ACT* is a true companion to this work, continuing the developmental process of linking the conceptual understanding of ACT principles and techniques to the fluid implementation of the therapy. As I will present moving forward, there is no arrival in a "final" mastery and outcome, but instead, an ongoing learning and engagement that brings ACT to life in the clinical setting.

The State of ACT Training

To date, how to best learn ACT, including its processes and intricacies, has not been well studied. A handful of articles exists, but they speak mainly to training in ACT for the purpose of testing its effectiveness. More research is undoubtedly needed. Despite the small number of studies, the research on training has promising outcomes. For instance, Strosahl, Hayes, Bergan, and Romano (1998) found that clients whose therapists were trained in ACT developed significantly better coping than clients of untrained therapists. This is good news. It's relieving to hear that training in ACT can produce that kind of outcome. The training involved an educational workshop with intensive clinical training and monthly supervision. Although these are effective modes of information transfer, the therapists' competency in the ACT model was not evaluated.

Research has also shown that novice therapists can be trained to adhere to the model in the absence of a treatment manual, and while doing so, produce a significant change in client outcomes (Lappalainen, Lehtonen, Skarp, Taubert, Ojanen, & Hayes, 2007). This is also good news. Another study found that training clinical psychology students in ACT has led to significant behavioral self-care changes linked to the core processes (Pakenham, 2015). This cheers me up. However, these studies did not evaluate the junior clinicians' competency or indicate whether it could be translated to the noted outcomes.

In one of the few studies of training in therapist competency in ACT (Walser, Karlin, Trockel, Mazina, & Taylor, 2013), following intensive training, clinicians reached only average competency as determined by their individual supervisors. In this program evaluation study, clinicians attended a three-day experiential workshop on delivery of the ACT theoretical model, focusing on the core processes and the commonly used exercises and metaphors. Following the training, trainees participated in six months of weekly, ninety-minute group supervision while implementing an ACT protocol with two clients. Trainees also audio-recorded ten of twelve treatment sessions, representing most of the implementation of the protocol. Supervisors listened to the recordings, rated them for skill and competence, and provided feedback to the clinician. Given the intensity of the training, it was somewhat surprising that on average the clinicians, as rated by their supervisors, were barely reaching average competence. However, according to the client outcomes measured in the study, this level of skill was sufficient for significant change.

It may be that the difference between a therapist who is doing ACT well, with competence and fluidity, and one who is struggling to do so matters less when it comes to clinical outcomes. A recent meta-analytic review of different therapies would support this supposition (Webb, DeRubeis, & Barber, 2010); according to this review, neither adherence nor competence played a significant role in determining symptom change. In the broader literature regarding these issues, the therapeutic alliance has proven to have a sizable impact on findings. Suffice it to say that it is difficult to fully know the role of competence in ACT in clinical outcome, let alone the training needed to get there, given the current state of the research. I would argue still that it is a worthwhile endeavor to distinguish between merely learning to apply the techniques and bringing that learning into practical implementation in a flexible, consistent, and heartfelt manner. Reducing ACT to a set of techniques that receive high fidelity scores reduces therapists to technicians and removes the heart from the intervention. Thus, there is much work to be done.

The Future of ACT Training

Many trained therapists note an ability to conduct exercises and metaphors without too much difficulty. The application of ACT techniques—delivering a metaphor or conducting an exercise—can be relatively easy and has seemed to enjoy significant success. Indeed, some of this success has led to an oversimplification of ACT wherein the techniques and exercises are viewed as the therapy itself. Other therapists and trainees have told me that they are excited about using ACT. They feel that they understand what is supposed to happen when using the model, they are eager to give it a try, albeit anxious, and then something very different unfolds when they enter the therapy room. ACT becomes stilted, inconsistent, nonresponsive to the client, or rehearsed, often leading the therapist back to a former way of doing therapy (pre-ACT) or to an overly simplified technique-based application of ACT. The latter reminds me of placing salve on itchy skin. It has the feeling of "laying something on top" of the individual seeking services, rather than working together with the client to bring therapy to its full interactive character.

Both beginning and seasoned therapists have shared with me their struggle to move past a technique-oriented quality of doing ACT and into an experiential and fluid implementation of ACT. It makes sense that a therapist learning a new intervention would focus on applying a technique. It is the movement to facile implementation wherein the challenge remains. Indeed, one of ACT's core training challenges lies in creating a way to communicate experiential work in the implementation of the core processes in a flexible way that enhances clinicians' understanding of all aspects of ACT (theory, technique, process, and heart) and its use with diverse and complex clients. Training clinicians to apply ACT flexibly and with competence goes far beyond describing exercises and how to implement them. It also requires connection to the real, in-the-room, and felt experience of clients who have varied presentations, problems, goals, and values. I believe it also requires connection to ACT at a personal level.

In my own exploration of ACT training designed to address the move from applying techniques to fluidly implementing processes, many considerations have come to light. First, workshops in ACT will not be enough to get you to fluid implementation and mastery. They are good; indeed, many are excellent, and they are a terrific starting point, but more in-depth work is needed. This can be seen in the questions I receive following training workshops. Clinicians have approached me following a demonstration or role-play and have asked, "How do you do it?" Simply responding, "Start by grappling with the underlying theory, contact the work experientially, know the story of ACT, and always keep the function of behavior in your sights as you work from a compassionate and heartfelt stance," just doesn't seem to suffice. I have worked to answer the question thoughtfully and thoroughly, yet I have noticed that the answer, although elegant in one fashion, doesn't necessarily assist the person in knowing what to do. Indeed, that response is often met, at first, with a look of excitement, rapidly followed by a look of confusion or deflation. The plain and simple truth is that to do ACT well and competently not only takes time in terms of grappling with theory, recognizing processes, and getting supervision and practice, it also includes understanding ACT from the inside out: connecting to personal flexibility by being open, aware, and engaged—both in your own life and in the therapy room. Although learning to implement ACT with competence and mastery can be challenging, it is quite fulfilling to help clients to connect to psychological flexibility as an ongoing process and to create meaning that can be engaged across a lifetime.

It is my hope that this book will begin to bridge this gap, assisting therapists in moving from fidelity to the model (the ability to adhere to the therapy, delivering it as designed using techniques and protocols) to competence (the ability to provide the intervention with proficiency and skill to the standard needed for it to achieve its expected effects) and perhaps even mastery (the ability to integrate a complex set of skills in an easily accessed, organized, and implemented fashion) with heart and wisdom (a stance or presence in therapy held by the clinician). This is no small affair and is ultimately about moving from ACT technique to ACT process, intention, and presence.

About This Book

As you can see by now, the intention of this book is not to focus on the basic detailed model of ACT or its underlying behavioral principles. These will be briefly explored, but from a kind of clinical-speak that is hopefully a shared way of talking with therapists in general about the model. My assumption is that readers already have the following understandings about ACT: (1) each of the processes in ACT is designed to create and support psychological and behavioral flexibility, (2) the essence of the therapy is working with clients and ourselves to engage in ongoing *acceptance* of internal experience while engaging in behavioral *commitments* that reflect chosen values, and (3) its clinical goals are exemplified in its very name and born out in its acronym: ACT.

However, the number of ways to implement acceptance and mindfulness strategies as well as commitment and behavior change strategies is vast. It is possible to guide clients in creative and new ways, untapped in manuals and protocols. The bellwether is consistency to the model. Are we guiding ourselves and our clients to be explicitly in contact with the two different ways of knowing the world—with the mind *and* with experience? Are we and our clients working to find a place where the stance taken in relationship to our experience (see McHugh & Stuart, 2012, for more on perspective taking) is such that painful thoughts, feelings, sensations, and memories are encountered in a nonjudgmental and open way? Are we and our clients finding the freedom to live by our values and not just the literal content of emotions and thoughts? Are we integrating the processes fluidly and tailoring them to our clients' needs and presentation, as well as to the functional aspects of behavior? Are we fully in the session—open, aware, and engaged—with heart? Each of these questions brings us to the purpose of this book—its raison d'être.

There are excellent books designed to assist you in learning ACT including *Acceptance and Commitment Therapy* (Hayes et al., 2012), *Learning ACT, 2nd Ed.* (Luoma et al., 2017) and *ACT Made Simple, 2nd Ed.* (Harris, 2019). Hayes, et al. (2012) is *the* foundational book and I hope is tattered and worn through reading and use for all wishing to learn ACT, its theory and principles. Other books for enhancing your therapeutic work such as fostering mindfulness in the therapeutic relationship (e.g., *Mindfulness for Two* [Wilson, 2009]) or combining ACT with other approaches like Compassion Focused Therapy (e.g., *The ACT Practitioners Guide in the Science of Compassion* [Tirch, Schoendorf & Silberstein, 2014]) are excellent reads for stretching your therapeutic proficiency. Two other books that are essential reads are *Mastering the Clinical Conversation* (Villatte, Villatte & Hayes, 2015) and, as mentioned, *Advanced ACT* (Westrup, 2014). Both are notable for their clinical wisdom, use of Relational Frame theory, and therapeutic examples. Indeed, the focus in each on experiential work and closing gaps between theory and practice have informed my work and are invaluable. Learning specific therapy processes as processes (e.g., Process-Based CBT [Hayes & Hofmann, 2018]) is also an essential read. This book, however, will focus on personal work with ACT, from the inside out, and on ACT as process—embedded inside of other processes such as clinical presence. Written in a nontechnical, straightforward fashion, it is designed to demonstrate the subtle and flexible application of ACT as a multilayered process and process-based intervention from a particular therapeutic stance.

Additionally, the therapeutic work done in ACT is enveloped in an overarching context of compassion for the human experience. Bringing behavioral principles to bear with warmth, genuineness, and a true sense of understanding for the client's circumstance is part of ACT's character. Compassion, sitting with and in pain, is part of what is called for in assisting clients to embrace ways of living that are meaningful and guided by values. The mind often interferes with this process. But not just the client's mind, the therapist's mind as well. The therapist is just as susceptible to fusion with mind as the client is, and may be at the highest risk for fusion in therapy when first implementing "the rules" or techniques of the ACT model. Therefore, as you read this book, I invite you to consider not just ACT's role in your client's life, but ACT's role in your own life and in your work as a therapist.

ACT as a Personal Journey

Even today, following years of experience conducting trainings, providing supervision, and doing clinical work in ACT, I haven't wavered from understanding that ACT is a personal journey. And through this journey, I have sought to share the possibilities in ACT with others suffering from painful pasts and unrelenting minds. This doesn't mean that ACT therapists are expected to walk around in a Zen-like or monkish state. It does mean practice. Openness and awareness are not easy in the presence of a busy and sometimes critical mind and in the context of full and often overwhelming lives. You will need to select this stance over and over again. Part of exploring ACT is recognizing that you are the place where experiencing happens, that you are invited to choose engagement, revisiting your values and creating meaning from the feet up, every day, every moment. Life is a process; it does not hold still. Therefore, as you read, you will be invited to routinely consider your own experience, to engage in personal reflections, and to ponder and digest the material. Your experience with ACT will not only inform how you do it with others, but also influence your therapeutic presence, your intention, your heart in ACT.

What Lies Ahead

As noted, the process work in ACT goes beyond the simple application of the model at the level of technique. Indeed, merely applying techniques moves the therapist away from the qualities of heart and wisdom that seem to so define ACT and make it a rich therapeutic experience. Rather, sessions are more likely stilted, disconnected, and nonresponsive to the client. If the effects of dissemination and oversimplification have brought a full tide cascading over ACT and drowning it in technique as therapy, then this book reflects an attempt to turn the tide back to two focal points of ACT: (1) its heart, or quality of benevolent openness to experience grounded inside of *being* itself, and (2) its wise implementation, or quality of pursuing self- and other awareness grounded inside of authentic interpersonal interaction. This kind of wisdom is linked to *being*, but sensitive to consequences and motion, recognizing that history participates, but that choice is critical. We connect to the moment, choose, and step forward and onward, ultimately until death.

The book is organized into two sections. Part 1, "Building Heart Through the ACT Core Processes," focuses on the three pillars of ACT—open, aware, engaged—which we will explore as personal processes. I will invite you to begin to reflect on your therapeutic stance in ACT, including your personal experience, and we will explore the concept of living life "from the feet up" (i.e., as a conscious choice). We will also examine knowledge of *self* and the relationship with *other*, existential issues and their impact on values, the challenges of personal control and letting go, and, lastly, what you say and how you say it in therapy.

Part 2, "Building Heart Through Unfolding Experience," will focus on layered processes in the broader context of therapy. Specifically, we will explore ongoing and overarching, interpersonal, and intrapersonal processes in ACT. Finally, in the last chapter we will turn to the ACT therapeutic stance—it's essence and necessity. A brief epilogue will bring the book to its end, but hopefully, it will be the beginning of a heartfelt and ongoing exploration for you as you continue your journey into the heart of ACT.

In developing the ideas for this book, I used a unique approach. I reviewed tapes of both ACT clinical sessions and ACT supervision, listening intently to the clinical scenarios presented and considering what ACT-consistent intervention might entail from multiple perspectives and the position of process. As well, I reviewed the questions in the supervision tapes that practitioners most commonly asked. During this review process, I noted the many complexities and nuances of ACT, as well as the various clinical problems and stuck points that therapists seem to get into when trying to implement ACT in different therapeutic settings and with diverse clients. It is these varied topics and my work with the fluid implementation and mastery of ACT that influenced and informed what I have presented in this book. There is a client-centered aspect as well: throughout the book I will convey how we hold and relate to the client as part of ACT's heart and wisdom.

The chapters that follow will cover many essential topics related to competency in the ACT model. In addition to discussing the model and exploring client examples, I'll suggest aspects of process and stance in therapy as part of the work needed to grow as an ACT clinician. At the end of each chapter you'll find a section called "The Heart of Now and the Door to Next," which contains a short summary of the main points of the chapter and a brief look at what's to come. My goal is to convey a level of knowledge that enables you to improve your flexible application of ACT and develop your therapeutic stance. However, an additional and essential goal is to convey a spirit of sensitivity, compassion, and heart toward you (the learner and clinician) and your work and to put into words a kind of wisdom that speaks to the clinical "presence" found in ACT.

As noted, the words "heart" and "wisdom," as used here, are meant to express a particular sentiment, a kind of showing up in this moment that is connected to experiential knowledge, not just verbal understanding. My intention is that my writing will reflect this quality, making for both a felt connection to the material and an accessible reading experience. As such, I will invite you to engage in many processes to enhance your experience with *The Heart of ACT*. First, I recommend that you audiotape or videotape a clinical session or two to best conduct some of the reflections (Reflective Practice exercises) you will be asked to do. Using a recorded session to evaluate and assess your own process at different times throughout the book will improve your experience of the book's intention. Additionally, you might consider reading the book and working through the Reflective Practice exercises with a colleague or during supervision. Asking and answering the questions in an interpersonal format will assist with the process. In each chapter you will find multiple Reflective Practice exercises, some inviting you to reflect on a taped session or interpersonal interaction, others inviting you to reflect on various aspects of your ACT practice. You can work through these reflections on your own, in supervision, or with a colleague or group of colleagues. Any of these approaches is valuable, but the interpersonal process may prove most useful. The Reflective Practice exercises are available for download at the website for this book, http://www.newharbinger.com/40392. (See the very back of this book for more details.) You may want to keep a journal to record your responses to these Reflective Practice exercises.

Finally, in each chapter, you will "hear" from two clinicians—Manuela O'Connell and Carlton Coulter—who have spent time learning and exploring ACT and have been grappling with its application with varied clients in multiple settings. Both clinicians are experienced in many therapeutic approaches and have come to ACT through workshops and other training. Their goals: competency, mastery, and compassionate delivery. In Manuela's and Carlton's reflections, you will read their reactions, personal experiences, and questions about ACT. Their reflections intend to provide a diverse view while expanding on the material for clarification and demonstration. I have supervised both Manuela and Carlton, and I am grateful for their contributions.

Ultimately, I hope that this book will provide the groundwork for a personal experience that sets you on the path, feet first, to the heart of ACT.

PART 1

Building Heart Through the ACT Core Processes

CHAPTER 1

Getting *Your* ACT Together

Knowing yourself is the beginning of all wisdom.

—Aristotle

Most people do not fully acknowledge or are reluctant to see that life contains suffering. But recognizing this truth is a first step on the path to transcending it. It is in this transcendence that vital and meaningful lives are born. As ACT therapists, we can assist our clients in this process by helping them to acknowledge and hold pain while continuing to take steps, each day, each moment, that are connected to personal meaning. This is done, as is often said in ACT, with head, hands, and heart. *Head* refers to verbal knowledge, the necessary intellectual understanding of the intervention. The work of the *hands* is about behavior; physical movement and taking action are fundamental. ACT's *heart*, however, appears to be more elusive. This is partly because it lies in the experiential nature of the therapy, not wholly in verbal understanding. It lies in the qualities of pure consciousness, in what it means to be an experiencing being—whole and alive. As the experiencer, one is observing and open to the flowing process of life, including its joys and pains. Inside this perspective, awareness of the fluid process of other is also present—that *all* human beings are whole, alive, and fundamentally acceptable. The therapist can connect to the client in this recognition, both being capable of transcending suffering across time and of creating purpose in an ongoing process of engagement in values.

Doing ACT well then is about getting *your* ACT together in a full and integrated way. This begins with personal work on being open, aware, and engaged. In this chapter and those that follow in part 1, I will invite you to explore your relationship to self and other through reflective and clinical material. This exploration, generally focused on the three pillars of ACT (open, aware, engaged), is intended to shape your personal growth and movement toward a masterful and heartfelt therapy.

The Big "Growing in ACT" Question

Have you ever had the experience of tasting a delicious dish when eating a meal out, and then, when trying to re-create it at home, it just wasn't the same? You simply weren't able to attain the right amount and kind of special ingredients that made the dish so pleasing to eat. Re-creating the dish might mean knowledge of the ingredients and close observation of the chef. But watching the chef cup spices in his hands or pinch them in his fingers and throw them in won't give you the exact measurements. Learning to spice the dish will take personal practice—it will take "getting a feel" for how much spice to add.

Listening to or observing a masterfully delivered ACT therapy session is like tasting a delicious dish and wanting to eat it again, but in your own "home." So, are there special ACT ingredients that lead to a masterful and heartfelt delivery? If so, what are they, and how much of each goes into a session? Or more specifically, "How does a therapist learning and growing in ACT get to the interaction *between* client, therapist, and intervention, occurring in a context, such that the function of behavior is targeted for change and principles are implemented in an ongoing and process-oriented fashion, and with heart?" I hope to explore this big question throughout this book in such fashion as to assist all clinicians wanting to grow their skill in ACT, but I also recognize the question as complex and challenging. Given ACT's experiential nature, the heart of ACT isn't easy to describe in words. However, by exploring personal engagement and clinical examples in ACT, and the processes involved in it as an experiential therapy, I will make a sincere effort to teach you the distinct ingredients that will enhance your skill in delivering ACT.

Developing the capacity to discern and deliver ACT, in an authentic fashion, as a multilevel, highly experiential process, will mean starting from the inside out. It will be about knowing the underlying theory and model, true, but it will also be about the personal experiential process itself. It is a felt sense and is about connecting to and using experiential awareness in deepening a flexible, client-centered, and process-based intervention. It is here that you are invited to start your personal journey exploring ACT's heart. It is a call to deliver ACT with intention and presence, connecting to your flexibility through being open, aware, and engaged in your personal life—and the therapy room.

Let's first visit three "prerequisite" suppositions about developing competency in ACT that will provide a kind of ambiance for the chapters that follow: (1) There is no arrival; (2) There are two kinds of knowing: reflecting and engaging; and (3) Discomfort precedes growth. Below we will explore these in detail.

There is no arrival.

Many therapists have asked, "What is the best way to learn ACT?" This is hard to answer. It seems there is no "best" way; there are many paths. You can start with reading and then move on to a workshop, or the reverse. You can start with supervision and move on to watching videos and talking with peers about application. You can start with peer consultation and move on to audiotape review, workshops, and reading. The point is that nearly anywhere is a good place to start; it's just that there is no good place to end. There is no arrival. Developing personal awareness as well as other awareness never stops. There is always a new moment, a new understanding or insight, a further action to take. Engagement in ACT as process means no arrival for you as a developing clinician and no arrival for the client who desires acceptance as an outcome.

One challenge in learning ACT is the wish to have an arrival—a final place where one is fully trained. This challenge has to do directly with the distinction between verbal knowledge and experiential knowledge (as well as the difference between process and outcome—ACT being a process-based therapy; see Hayes & Hofmann, 2018). Both are important, and neither should be sacrificed in the service of the other. Verbal knowledge can give the impression that after years of study, an arrival is imminent. You have read all that there is to read, attended the requisite workshops, practiced with supervision. However, contact with the experiential world, getting ACT from this perspective, having this kind of knowledge, tells us that arrival isn't imminent. Indeed it tells us there is none. Still, I will suggest that ongoing connection to experiential knowledge is a must. This work is harder to do, especially without the support of a formal supervisor and ongoing self-awareness work. This difficulty may be the reason that many therapists fall back on discursive learning such as reading and workshops. Like our clients, contacting the experiential world can be painful. But it can also be freeing.

There are two kinds of knowing: reflecting and engaging.

Reflecting on what it means to be human is often fraught with "mind-y" qualities that are logical and socially designed. A good life is often presented as one that is linear in nature (e.g., life events will unfold in a particular way) and full of wanted outcomes that are orderly in arrival. We will have a happy childhood, followed by a happy marriage with children, a work-filled and fulfilling adulthood, and a happy retirement, followed by a certain, but largely unwanted, death. The problem with this formulation (or whatever formulation one has) is that life is not orderly. When we stop and rest in what it means to be human, to be alive, when we touch experience outside of what is socially designed, we learn that life contains a bit of chaos, both inside and outside of the skin. Our experiences, both personal and shared, include simplicity and wanted outcomes, complexity and disappointment, gain and loss; life is filled with amazement and fear, joy and pain, belonging and loneliness. When we let ourselves connect to life as experienced, we become conscious of the realities it contains—it is fraught with struggle, and it is rich with wonder. Ultimately, this connection, this awareness of experience, also points to life's brevity—an urgent sorrow pleading with us to engage.

Engaging, however, means recognizing that life is more than what our mind makes of it. Once we begin to participate in language fully, we start to lose contact with experiential knowledge and fall away from this kind of wisdom. Speaking more technically, we begin to live by rules (rule-governed behavior) and move away from what we have learned through experience (contingency-shaped behavior). It is in this shift that perhaps rigid fidelity to ACT "rules" and structures takes form. The wisdom of experience is enveloped by sense-making and leads therapists down paths of explanation and understanding. Bringing ourselves back to experiential knowledge is not only central to well-being (living well as a being), it also has something to say about what it means to be an ACT therapist.

When we contact experiential knowledge in the moment, it tells us something about *being*. We learn that the internal events are rising and falling in a continuous stream of experience. It is here that another kind of richness in life is contacted; it is felt experience in the moment (as well as learned experience from the past). As ACT therapists, we join our clients, viewing them (and ourselves) as a context for ongoing experiential events that include all things sensed, thought, and felt. Moving into the space of conscious and connected well-being means encountering life and its inherent chaos, both internal and external, with flexibility—being able to respond, in the moment, in a way that is healthy and values-driven. It will be challenging to communicate about and connect to a client's experiential work if you, as a therapist, have not done it yourself. An awakening to your experience is part of "knowing" ACT. This doesn't mean that you *must* have a particular experiential moment during a workshop or arrive at an enlightened space while learning. Indeed, feeling forced to do these things misses the point. It does mean contacting experience, recognizing its fluid nature from a broad and felt perspective. It means sensing, feeling, thinking; being aware of each of these and their ultimate transience. It means showing up to personal pain from a conscious position, paying heed to its fleeting character as well as attending to joy and its ongoing or ever-present rise and fall—there is no arrival here either. This is the process of being alive and, I would submit, part of the heart of ACT.

> ## Manuela's Reflection
>
> *Why do you think contacting an experiential sense of self is a must? In my own experience as a developing ACT therapist, I quickly discerned the importance of the two ways of knowing, through reading and understanding the approach theoretically and through experiential knowledge, in order to connect to the work from a felt and personal level. To gain a deeper understanding of the theoretical knowledge, I investigated and explored each of the ACT processes through the lenses of my own experience. As well, I watched others who seemed to embody the approach—knowing ACT with their heads and living it in their lives. In my supervision work, the experiential side of ACT was evoked in process and procedure. I was learning from my own experience that focusing on the experiential nature of ACT was essential in my training. It seems that many therapists are not used to doing therapy in such an experiential way because they are taught to follow certain rules to do ACT right.*
>
> Robyn: *I suggest this as a must for several reasons. It is easier to model and communicate experiential knowledge if you are personally aware of what it is. If a clinician is unable to recognize nonverbal experiencing and learning, it will be difficult to convey it to a client. As noted, I am not suggesting that therapists need to have an "awakening" or even a particularly powerful emotional experience during training. Rather, the capacity to observe that humans are more than minds and to connect to the intrapersonal, emotional, and sensing part of ourselves is vital. Willingness to experience in this fashion not only provides a model for our clients, but it also allows for genuine and authentic emotional exchange.*

Discomfort precedes growth.

As time has passed and ACT has grown and succeeded, another issue has begun to show itself. Dissemination of ACT has been broad and supported. It is a good thing for ACT technology (the methods and procedures) and training to be transmitted around the world. There is a risk, however.

Along with dissemination comes a kind of dilution—the therapy can be viewed as a set of clever exercises and metaphors. Process and therapeutic stance can be lost to the execution of techniques. That is, the techniques of ACT become the therapy itself, as opposed to a therapy that uses techniques or methods and procedures. This problem has the potential to lead therapists in the opposite direction of competence and mastery, away from process and function. More ACT techniques do not necessarily make a better ACT therapist. However, the techniques are "easier" in that you *can* arrive at a destination.

As I reflected on the content of this chapter, I wanted to note this growing concern. I wondered if pointing out that technique as therapy is problematic might be experienced by some as challenging, confronting, or perhaps even controversial. I am suggesting here that knowing the methods of ACT, even if you know them well and understand what process they speak to, is just not enough. The heart of ACT is not in its techniques.

As therapists and human beings, we get attached, myself included. I have individual styles, preferences, and favorite ways of conducting therapy that I cling to. Inside of this attachment, I often don't *really* want to hear dissent or be questioned or tested concerning my affections. It is part of why I invited Carlton and Manuela to participate in the writing of this book. They have each pushed me to consider alternatives to my ideas and have questioned my firmly held stances. They have assisted me not only to speak more clearly to my theoretical orientation to therapy but also to look at, appreciate, and articulate with coherence and depth my functional contextualistic approach—that growth is about ACT process and relationships, not simply the use of ACT techniques. As well, they assisted me in letting go when I was too tightly attached to any single idea about the way ACT should be implemented; there are many ways as long as the implementation is theoretically consistent. So, I invite challenge and confrontation to my ACT work. It is in this space that possibility for change arises. Discomfort precedes growth.

Manuela's Reflection

I remember well the first two supervision experiences with Robyn after attending multiple ACT workshops. The first thing she asked me to do was to follow a protocol. I came back to the second meeting embarrassed because I wasn't able to do it. I could see that my client's needs didn't fit the first session of the protocol. I was trying to convey session one of a protocol about creative hopelessness, while all the cues in the session were about values. I was struggling with the tension between what I was supposed to do, follow ACT rules, and being present and engaged, contingency sensitive. I still experience that tension in my practice, and I aim for flexible equilibrium (the balance between following protocols and reading in-the-moment experience) as part of my work in mastering ACT.

My hope is that, like Manuela, both seasoned and novice ACT therapists will reflect on how they implement ACT with clients and how they train others in the approach. Questions to ask might include:

- Have techniques begun to overtake process?

- Has the therapist's original hope of serving others by means of this model been diminished by any kind of narrowness in focus?

- Have new, old, or otherwise used ACT interventions become "slick" (i.e., used as a clever way to conduct the therapy)?

- Has functional analysis of behavior been lost to simple procedure?

- Is the therapist clinging to metaphors and exercises as the therapy, sacrificing flexibility and relationship?

When reflecting on your own ACT work or reviewing your sessions across time, it will help to be aware of how much and when you are implementing techniques and calling it ACT. You'll want to not only ask yourself if you understand their purpose, but also assess how you explore or process exercises in an ACT-consistent way. More importantly and to the point of this section, if you find in your review a place where you are engaging the techniques as the therapy, not entirely pulling them together in an overall approach grounded in case conceptualization, ACT theory, and ongoing process, remember that discomfort precedes growth.

Carlton's Reflection

Robyn, in this chapter you have identified some of the common barriers to effective ACT therapy, including the therapist focusing on technique at the expense of the process, and following rules at the expense of experiential knowledge. When I reflect on my own therapeutic practice, I clearly see that I have run into both types of problem, particularly when first starting to work within an ACT approach, but also subsequently when I should have known, and actually did know, better. I would anticipate that these two problems are common among ACT therapists. This then begs the question, "Why as therapists are we so prone to this?" For me, some of the answers to this question exist in what I experience before I even enter the therapy room; for example, I really want the therapy to be effective, and in order for that to happen, I can sometimes drive too hard toward change and lose focus on experiential knowledge, process, and flexibility. Further answers to this question are found in what I experience once I am actually sitting opposite the client; particular presentations and client behaviors elicit reactions in me that I can sometimes respond to with either the indiscriminate use of technique or recourse to a rule. Developing awareness of intra- and interpersonal factors (topics we turn to later) can be facilitated by some of the exercises included in this book, through personal reflection, and in the context of supervision.

However, there is another critical source of influence on these therapist behaviors, namely the way that ACT is often disseminated. Pick up almost any ACT book, and you won't have to turn too many pages before you find examples of "experiential exercises," and workshops often have a very similar focus. It seems that the experiential nature of the ACT approach is all too frequently translated directly into both set pieces and practical techniques that are subsequently to be "used" by the therapist. It's little wonder then that therapists coming to the approach learn that those exercises are what ACT actually is.

Furthermore, I have also come to notice that in books, articles, workshops, and conversations with other very experienced ACT therapists, some of the theory of ACT has become enshrined into rules, often stated as absolutes; mind is bad, control never works, one must always accept. Sometimes these messages are implicit; at other times they are explicit. Functional analysis of the client's actual behavior in context, or indeed analysis of any kind, is replaced with the dogmatic imposition of rules. The therapist's job then becomes to make the client realize that this way of looking at the world is the truth, and the best way for the therapist to achieve this is to throw more and more technique at the client until they finally submit to this new worldview. I agree with you that one way of addressing the problem of technique is to focus more on awareness of intra- and interpersonal experiential processes, to bring more "heart" into the therapy room, but I also think a greater emphasis on analysis, functional and otherwise, is required to guard against the imposition of absolute statements and judgments about the behavior of human beings and the factors that influence it. This would, however, mean that the therapist must be willing to acknowledge that sometimes mind is by no means all bad, that control can work very well, and that acceptance is just one possible way of responding. Would you agree?

Robyn: This is an interesting observation. Are therapists relying on technique instead of process because this is the way ACT is being taught? There are probably times when dogma and rule following are happening in training. I know my passion during training has the potential to be interpreted that way. It is probably wise for all who supervise and train others in ACT to reflect on their training process and for clinicians to reflect on their clinical work. And, yes, I would agree. Your words speak for themselves, so I will simply add: functional analysis is imperative. Mind is not all

> *bad, nor is control a problem unless it is misapplied, excessively so, to internal experience in such a way as to cause problems for the individual.*

We've now come to our first Reflective Practice exercise. This, along with all of the Reflective Practice exercises in the book, is available for download on the companion website for this book, http://www.newharbinger.com/40392.

REFLECTIVE Practice 1.1

It might be useful to routinely engage in self-assessment of your use of ACT techniques and procedures. Or you might explore with colleagues the nature of the ACT therapeutic work you do. Reflect on whether you need to step into discomfort to grow your ACT work. I invite you to ask yourself the following questions:

- How often and when do I use ACT techniques (e.g., standard ACT metaphors and exercises) in session?
- Am I open and thoughtful, or do I have an agenda that I feel my client(s) must adopt? Look deeply before answering this question.
- Does that agenda include a set of techniques I routinely use? Have they become rote?
- Is my enthusiasm for ACT tempered by interventions that include choice making and an overall process-oriented approach to human suffering? Do I convey that acceptance is a choice? Do I note that control can be useful depending on the context?
- Do I recognize my own experience as process and bring it to the therapy in a way that is functional?
- How do I feel about certain exercises? Are there ones that I avoid or ones that I use too often?
- How could my use of ACT techniques potentially influence my overall ACT work and my relationship with my clients?

Now that you've taken a close look at your own approach to doing ACT, let's look at some of the pitfalls that might be hindering your growth in ACT.

Roadblocks to Growth

Two barriers to developing competency in ACT have to do with the overuse of techniques over process and the excessive involvement of mind. Let's explore each of these in depth.

Overuse of Technique

When we say, "technique as therapy," we mean that the therapist is merely relying on the techniques as their way of doing ACT. Overuse of technique can take many forms. Holding too tightly to a newly learned or favored intervention can serve to shut down the functional assessment process. You begin to search for *how* to use the technique rather than looking to see if the technique *should* be used. For instance, you might alternate between holding a clipboard or piece of paper up to the face and away from the face to demonstrate defusion every time a client is fused. Or you might focus too heavily on the paperwork of the intervention (e.g., homework, metaphorical images or tools) by routinely placing a piece of paper between the client and yourself. In so doing, the interpersonal dynamic or emotional experience is lost.

Other examples of overusing technique include frequently asking the client how they feel "right now" in the present moment during a single session or reducing self-as-context to the chessboard metaphor. These methods may "qualify" as ACT-consistent work, but used simplistically, they can miss the more substantial interpersonal, experiential, and process-oriented part of the work. The therapy can begin to look a bit cartoonish, and the very work of altering behavior in a flexible and responsive style may be given over to excessive technique-oriented therapy. This can cause problems including avoiding experiential work, focusing on form over function, promoting fidelity rather than competence, and attending to content, not process. When you become overly invested in or devoted to such techniques, you can end up curtailing clinical judgment, applying techniques in a noncontingent way, or having the client's individual needs disappear under your maneuvering with the ACT tools.

Below, we will consider two specific techniques to help illustrate the issue.

THE "WELL NOTICED" TECHNIQUE

A quite viable technique in ACT is a short statement designed to assist the client in the practice of "noticing." The goal is to reinforce noticing itself, not necessarily the thing noticed. So, for example, when a client states that he "noticed feeling upset," the therapist responds, "Well noticed." There is nothing inherently wrong with this technique. Indeed, it might be quite fitting given the context or issue at hand and may serve to reinforce noticing behavior. However, it can be overused or used in a way that is experienced as aversive. If you simply say, "Well noticed" without assessing its function or how it might "land" with your client, it is likely that you won't know if you are actually reinforcing noticing or punishing it. Suppose a therapist truly likes this technique and starts to use it on a regular basis, but does so without vigilance as to "the when and where of using it," without attention to its function. Imagine the following conversation between a client and therapist:

Conversation 1

Therapist: How did your homework go this week?

Client: Well, I didn't do it all the time, but there *were* times when I was noticing that I felt anxious when my boyfriend was talking to one of his friends… she's a girl. I didn't think it would bother me, but it did and still does.

Therapist: Well noticed.

Here, the therapist has reinforced the noticing itself and doesn't necessarily need to respond immediately to the content, anxiety, and issue of the boyfriend talking to a girl. Depending on what the therapist was trying to do, reinforcing noticing might well be a fine intervention. The noticing behavior can be explored first, and the anxiety can be worked with later in the session. Let's look at another example.

Conversation 2

Therapist: How did your homework go last week?

Client: I have been wanting to talk to you about how things are going. I don't think this stuff works for me and I am not sure about this [*pointing rapidly back and forth between herself and the therapist*]. I couldn't really do the homework. If I noticed anything, it was that I am not liking how things are going. I am feeling bad about saying anything.

Therapist: Well noticed.

Using "well noticed" at a time when the client is questioning the therapeutic relationship and therapy is potentially doing it for its own sake. The therapist in this scenario has not only dismissed the client, but it also seems she has misunderstood the use of the intervention. She has followed a rule (i.e., say "well noticed" when the client notices something) rather than considering the function of the intervention. At a time when a client is expressing this kind of vulnerability, it is important to be present to the moment of the experience and to tune into the interpersonal dynamic. It may not be the best time for using this technique. It may not be functionally responsive. Dismissing clients with "well noticed" may leave the client feeling lost or even angry. The technique becomes aversive.

Problematic uses of "well noticed" may include stating it (a) in a noncontingent way as a canned response, (b) because it is linked to a homework assignment, (c) because you heard it in a workshop or read it in a book, (d) as a content-oriented and not necessarily functional or process-oriented response, or (e) even more problematically, because it seems clever or is viewed as a "smart" intervention. Rather, if the therapist is attending to function, then Conversation 2 might unfold differently.

The therapist would change direction based on the functional assessment occurring in the moment (e.g., What is the function of the client stating that the therapy isn't working?). If the therapist is aware that this is an avoidance move on the part of the client, then she might initiate a conversation about the problem of control, exploring whether the client is stepping back from therapy to escape emotional pain. Or if the client doesn't speak up in relationships, the therapist might say something about how grateful she is that the client was able to speak about the therapy and the relationship, moving to create an open and honest dialogue about what isn't working for the client. The therapist changes direction, letting go of the technique—"well noticed"—and instead meeting the needs of the client in that moment.

WILLINGNESS AS THE ALTERNATIVE TO CONTROL

A second example of technique as therapy involves the introduction of willingness as the alternative to control, illustrated in the example below, pulled from an audiotaped session:

Therapist: [*after introducing willingness as the alternative to control*] There is this exercise we need to do now that is kind of weird.

Client: Um, okay.

Therapist: So, we are going to be willing, and what we do is look into each other's eyes for a period of time while remaining silent. Are you willing to do that?

Client: I guess so.

Therapist: Okay, I will start the clock and let you know when we are done. Okay. Here we go [*Therapist begins looking into the client's eyes, and the client looks back. This process is engaged in for about forty seconds*].

Therapist: How did that go?

Client: Um…okay, uncomfortable. Why did we do that?

Therapist: It is about willingness. And, yes, super uncomfortable. There is another exercise we can do though…

To many therapists, the problem will be immediately apparent in this example. The exercise is nearly wholly decontextualized and appears to be done as part of supporting willingness but falls flat in its delivery and purpose. The most challenging part about the example for me is that the therapist is someone I had trained in a workshop. The best part about the example is its teachable qualities, for the therapist and me. We were able to explore what was happening to him and look at personal willingness. The therapist in this interaction loved the experiential nature of ACT workshops and was moved by the "eyes on" exercise (Hayes et al., 2012) in particular. Fully excited about using ACT therapeutically and keen to try new things, he was highly motivated. However, when it came to implementing the exercise in session, he became afraid. He got locked into personal and social evaluation, was anxious, and pulled back: *People don't do this. I can't remember what to say. What do I do? This is stupid. I don't want Jeff (the client) to see how anxious I am. Please, let me just get through it.* The therapist, unwilling himself, struggled and then rushed the exercise to avoid his experience. The technique became the intervention.

THE ANTIDOTE: BOOKENDING TECHNIQUES

If you find yourself in these kinds of predicaments for whatever reason—for instance, you missed something in the training or had trouble absorbing the point of the exercise—keep this in mind: if exercises aren't "bookended" (i.e., contextualized), then they will seem to be done for their own sake. The value of bookending exercises can't be underestimated. Knowing why you are doing an exercise is more important than getting the exercise right. Knowing the purpose of an exercise is not about understanding merely where it fits in the model, but also how it supports or advances a process given a particular client issue. Additionally, being aware of your presence and state in delivering an exercise will also be important, in case you might rush, skip, or gloss over it to avoid anxiety or judgment. If you don't know the purpose, at this moment with this client, it is better to leave the exercise undone. Or you can work with the client. Speak to your own anxiety, letting the client know your hypothesis. Work together to bring the exercise to life, as in the following dialogue:

Therapist: [*after introducing willingness as the alternative to control*] There is an exercise we can do to experiment with willingness. I have thoughts that it is odd, but I think it has something to show us. Both of us. Are you willing to be curious with me?

Client: Um, okay.

Therapist: So, we are going to engage willingness, and what we do is look into each other's eyes for a period of time while remaining silent. Are you willing to do that?

Client: I guess so.

Therapist: Okay, I will start the clock and let you know when we are done. Okay. Here we go [*Therapist begins looking into the client's eyes, and the client looks back, for about forty seconds*].

Therapist: Okay, let's explore. I will speak to what my curiosity had to say about the exercise, and I would like to hear yours. Mine said it was unusual, and I was feeling anxious. What did yours say?

From this place, the therapist and client can unpack an exercise together, exploring its purpose, noting how willingness to do it was in the doing of it, not in the getting it right or in avoiding the "oddness" of it. It is important to be aware of your own process in the therapy as you begin to experiment with delivering exercises and metaphors. Own the techniques as part of a journey in therapy, rather than the techniques owning you. Be curious, open up, and see what unfolds. Notably, in the initial example given, the therapist was following an ACT protocol, and therefore the exercise was called for during the session. A high score for fidelity to the treatment protocol might be achieved—the exercise was indeed done. Competency, however, is entirely at risk. It might also be worth noting that this was a seasoned therapist; we are all susceptible to bumps on the journey. It is the capacity to be flexible in your use of exercises and metaphors—to shift and change according to context and function—that will assist you in creating a therapy that uses techniques, not the other way around.

Manuela's Reflection

I remember the excitement when I first started ACT. The workshops were emotionally moving, and at times I felt I was on a roller coaster. In my first workshop with Dr. Steven Hayes, there was so much happening. I learned a new theory that was counterintuitive, I did a lot of personal emotional processing…and then I had to go back to life and to my professional setting. I remember vividly having this thought running inside of my head: And now what? There I was after the workshop, baffled and unsure. How do I do ACT? How do I do this with my clients? With trepidation, I approached Dr. Hayes (the developer of ACT) and asked him how I was supposed to do ACT. His answer will always accompany my training journey: "Let ACT begin to be embedded in your own personal practice." At that time, I didn't fully understand the multiple layers of the advice. I became concerned with learning every technique and making my mouth say "ACT words." Doing ACT interventions "properly" with a protocol seemed paramount. It wasn't long before the vitality and purpose of that first workshop began to drain from me. I was more dedicated to doing it "right" than to working with the client. My therapy sessions were more like a performance. I knew something was missing. I turned to supervision and began to connect to the many layers of learning ACT. ACT began to be embedded in my practice in a truly genuine way.

The Excessive Involvement of Mind

How is it that we get stuck in technique? Let's return to a fundamental concept in ACT. Much of the psychological suffering of humanity is related to the mind's excessive involvement in experience—minds go to work trying to control, manipulate, manage, undo, stop, and otherwise fix or understand. Excessive use of mind emerges not only in the therapist's life but also in the therapist's sessions with clients and the therapist's work to understand an intervention. Behavioral rigidity that emerges from our rules—our mind—comes in many forms and can lead to problematic relationships with our own experience as well as with others. For the therapist practicing ACT, getting it right, doing it smartly or cleverly, or rigidly following a protocol may all be part of the loss of flexibility. Not being willing to be vulnerable, anxious, sad, or tearful; not being willing to be with ambiguity, distress, and discomfort; or not being willing to feel what one feels in the space of silence can all hinder flexibility. It is here that I hope to stretch your understanding of what it means to do ACT as you move through the chapters in this book. No longer will you be filled with negative and positive emotions as defined by mind and sorted behind the social world and medical approach to human suffering. Preferably, it is about you and your client standing in a place where you are filled with a rich and varied emotional life containing all nature of feeling, thought, and sensation, each connected to living meaningfully in the world, and in your therapy sessions.

Letting Go: Opening to Change

As we have seen, letting go of technique as therapy doesn't mean that you shouldn't use ACT exercises and metaphors. Bringing them into the therapy is part of the work. However, one of the core competencies within the ACT therapeutic relationship is avoiding the use of "canned" ACT interventions, instead fitting interventions to the particular needs of particular clients. The most important part of this ACT core competency is *changing course to fit the needs of the client at any moment*, and the most important word in this competency is "changing." Below are strategies to assist with this process.

Balancing Client, Therapist, and Intervention

It may be helpful to rebalance the therapy work in any efforts you make to step away from overreliance on technique. It is important to consider the ongoing process of the therapeutic relationship in the arc of therapy. The therapist, client, and intervention are all part of the process, as shown in the diagram that follows.

```
           Client


Therapist         Intervention
```

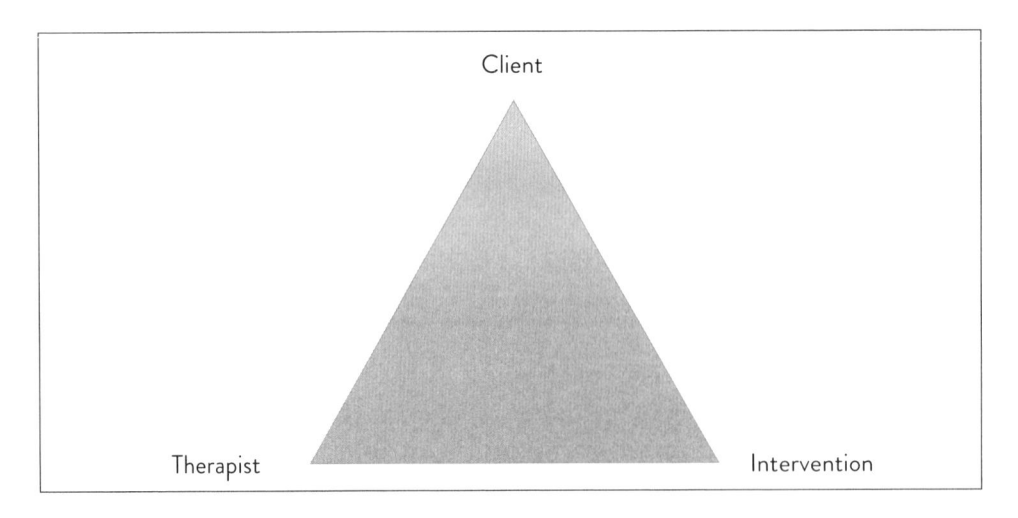

Striking a balance among these three elements across time can assist the therapist in deeming whether too much time or energy is being spent in one of these three "areas" over the distance of the full therapy. That is, if the therapist is frequently talking, and is "overemphasized" regarding what is happening across time, then a rebalance is needed. The same is true for the client, by the way. If the client is doing the majority of the talking over time, then a rebalance is worthwhile. An overfocus on intervention (i.e., the therapy) can also occur. This can take place when therapist and client spend large amounts of time looking at pieces of ACT paperwork that have been placed between them, or when both are engaging in analysis, explanation, and understanding of ACT. This can also happen when the therapy is packed full of exercises, and the client doesn't get the chance to talk or engage with the therapist on other levels. Again, a rebalance is needed. Maintaining a reasonable balance between therapist, client, and intervention over time can assist you in many ways. For instance, it may help you to consider your own presence in therapy (e.g., *I am talking, explaining too much*). Or it may help with recognizing if the client is caught in storytelling (e.g., *Is the client always talking, explaining their behavior?*). It may help you to step back from a protocol or favorite tool when you're focusing on it too much, and the relationship is getting lost to the tools and techniques of ACT. Keeping the triangle of therapy in reasonable balance over time is a lightly held self-check on the process.

Setting a Therapeutic Intention

Setting a therapeutic intention involves choosing a stance toward the client and then consistently responding from that stance. This stance is a personally selected position consistent with behavioral commitment and most broadly linked to self-as-context. It is one of holding the client as fully capable and whole. I will define and explore the therapeutic stance more fully in chapter 10 but will briefly note some of its qualities here.

Inside of the ACT therapeutic stance, it is essential to be mindful of the kind of languaging that therapists can get caught up in concerning their clients. At times, therapists will find themselves talking about clients with others, talking to clients themselves, or thinking to themselves specific thoughts about the ability or capacity of a client. I have noticed a tentativeness that can be projected in these dialogues. It manifests in holding the client as fragile or incapable. The problem with this perspective is at least twofold. First, and most importantly theoretically speaking, it undoes what should be a very defused, noticing language as language process that ACT holds. If I buy that the client is fragile and incapable, then I am fused with a concept about the client. I have moved away from the perspective that the client is a larger whole—a being that "holds" experience—and I have bought the idea that the client is inherently broken in some way. It makes sense then, to remain theoretically consistent, to hold these ideas of client as fragile and incapable lightly. Standing firmly in the position, session after session, that the client is whole and capable is about recognizing the experiencer and their ability to respond, respectively.

Carlton's Reflection

Isn't taking the position of holding the client as whole and capable also being fused with a concept about the client?

Robyn: *It could be. Fusion occurs when there is no awareness. In choosing this stance, there is an awareness that is being selected as a stance. There is a recognition of the position. It is not "mind-y" and tethered to expectations. Rather, it is defused from holding the client as broken and is a way to recognize their larger experiencing sense of self as well as supporting behavioral change. The client is able to respond and create something different in their lives.*

Holding this stance encapsulates, as well, what it means to know that change is possible, even for clients who have incredible pain and what appear to be insurmountable obstacles in their lives. I have worked with therapists in supervision who state, "The client is too fragile; I don't think I can say that." It is essential to understand what the therapist means by this statement and what actions he or she is taking in response to them. Often, the therapist is feeling afraid or worried while also letting these experiences dictate their actions in therapy. The client will hear or sense this, and the potential for movement forward will be slowed. This is not to say that there may not be functional reasons for not doing or saying something in therapy, but it might be wise to be aware of the difference. Intention, self-knowledge, and the function of your behavior with the client are essential in this place. Clients held as fragile from a fused place may not be told or given the help they need. Additionally, not holding a stance that the client is whole and capable seems to refute the overall objective of therapy itself—assisting the client, as defined by him or her, in making positive change.

The stance of whole and capable includes a complete openness to the client's struggle while holding a firm position that change is possible. This stance "feeds" the interpersonal relationship in an ongoing way. There is nothing the client can say or feel that undoes the stance; it is a stance of love for the client as they are and genuine faith in their capacity to behave in ways that are values-based. If the therapist is present and open in this way—holding the client as whole no matter what they feel, think, or sense, and as capable of change, even in the face of great difficulty—it is likely that they will experience a process of interpersonal connection with the client that includes respect, deep listening, and mutual empathy inside of a safely vulnerable and developing relationship.

> ## Manuela's Reflection
>
> *What would be the way to work on this stance—holding the person as capable and whole? As I see in my own practice and in that of my students, it can be easy to get caught up in descriptions of the client as well as their own descriptions of themselves as broken, faulty, or incapable.*
>
> Robyn: *First, it would be helpful to defuse, in session, when you become aware that these evaluations of your client are arising. However, in many contexts, including when sitting in front of a client who claims they are broken or when working in a system that holds people as disordered, the process of defusion may seem inaccessible. This work involves coming back to the recognition, as noted in the introduction to this chapter, that clients are experiencing beings, just as we are, and that "broken," "faulty," and "incapable" are evaluations that keep people stuck. Second, working on the stance also includes your presence in the therapeutic relationship. We know that authenticity, genuineness, and unconditional caring for the client are important variables. I might also agree with Banks (2016) that our natural state is one of interconnectedness; we are hardwired to connect, to join—one human with another, both caught in language, but not defined by it. Third, the six core processes in ACT provide a model, and engaging this stance involves the exploration of these processes from a personal perspective, thus experiencing "self" beyond the labels.*

Integrating and Embedding ACT Technique

In support of ACT techniques and exercises, I have and will continue to argue that these are part of the therapeutic relationship process. However, they are fully embedded inside the therapeutic context, which includes you as well as the client. Indeed, ACT exercises can be said to lose their impact if they are separated from the therapist employing them and from the client with whom they are being used. As I have emphasized, the techniques and exercises are not the central focus; they are firmly enveloped in, or incorporated as a part of, a more substantial ongoing interaction. Therapist use of techniques and exercises, when embraced as part of a larger whole, is guided by a kind of listening "beneath" the surface of behavior (a concept we will explore further in chapter 2). For instance, when listening to a client, the therapist's responses, followed by use of ACT techniques, are based on more than a simple reflection of what the client is saying and what appears to be the story at hand. Rather, what happens next is based on a deeper kind of listening, a listening that "hears" the function of behavior(s), in this moment, occurring with this client, with their history, in this context. This is a listening that is broad in nature. It includes awareness of the experiential state of the client, the experiential state of the relationship, and the experiential state of the self. It takes in emotion, sensation, story, body language, history of interaction, current interaction, and intention linked to a direction moving forward. It is conscious awareness of self and other in relationship. It is mindfulness connected to workable action.

REFLECTIVE Practice 1.2

Take a little time to review your ACT clinical work. Take time to connect to how you are currently implementing ACT with your clients. I invite you to answer the following questions:

- Are there places in my ACT work that feel stilted or old?
- Are there elements or processes of the intervention that I tend to shy away from?
- Where does my mind capture me and steer me and thus my clients in unhelpful directions?
- Where do I think personal growth is needed in my ACT work?

Notice any resistance you have to looking closely at how you implement ACT. Consider an open, aware, and engaged stance as you move forward. Simply explore, and if willing, commit to the discomfort of change.

The Heart of Now and the Door to Next

The competent ACT therapist needs to master a complex set of skills, cultivating an awareness of multiple processes coinciding and coming together with the client in a shared understanding linked to purposeful activity. Recognizing the truth of no arrival, concerning both acceptance and experience (it is always there to be done and felt), means there is still more personal awareness and growth in knowledge to do. Observing, tracking, and attending to the overarching process of therapy across time, the interpersonal process, and the intrapersonal process also allows the therapist to contact, understand, and intervene at a functional level. Knowing your own patterns of behavior and in-the-moment responses, the client's patterns of behavior and in-the-moment responses, and the interplay between the two supports your capacity to have a responsive and flexible ACT practice. We will explore all of these issues in part 2 of this book. For now, keep in mind that practicing ACT is a lifelong personal endeavor as well as a lifelong learning endeavor. Start with learning ACT technique and protocol, but don't stop there. Use your personal ACT work to inform your experiential work in therapy. Keep stepping into the heart of ACT as you move on to chapter 2, where you will more closely examine the three pillars of ACT in your own life and therapeutic work.

CHAPTER 2

Living Life from the Feet Up: Open, Aware, Engaged

Infuse your life with action. Don't wait for it to happen. Make it happen. Make your future. Make your own hope. Make your own love.

—Bradley Whitford

Living life from the feet up is fully possible using the three pillars of ACT: openness, awareness, and engagement. These pillars set the context for an important shift in therapy. I often say to clients, "Our work will be about living life from the feet up, not the head down." In ACT, we invite the client, in varying and ongoing ways, to move forward in life, noticing mind, connecting to heart, and moving the feet. Aware and open is the antidote to "frozen" or immovable feet (stillness, inaction) for you or your clients. Thus, weaving the processes inside the pillars (e.g., willingness, defusion) into a fluid mix in session, linking and smoothly transitioning between each, is important, and is also part of a competence and mastery endeavor. Additionally, when adeptly done, the transition between processes has an effortless and nimble quality, meaning that the therapist has enough facility with the overall model such that moving between these processes doesn't feel forced or stilted. The work is elegant in form and manner and is ultimately and easily linked to the functional outcomes, done with the feet, outlined by therapist and client.

In this chapter, we will explore the three pillars of ACT—open, aware, and engaged—linking them to how your personal experience as a therapist and human being are part of what it means to do ACT well, for others and for yourself. I will invite you to explore your relationship with each pillar but will not spend time describing the processes other than to give a brief definition for orientation purposes. As noted, the ACT processes and clinical relational frame theory (RFT) have been well described in other books (e.g., Hayes, Barnes-Holmes, & Roche, 2001; Luoma, et al., 2017; Törneke, 2010; Villatte, Villatte, & Hayes, 2015). I will point to how the three pillars and the processes inside each, as well as personal psychological flexibility, play a role in your work as an ACT therapist.

Exploring Open, Aware, and Engaged

Becoming adept as an ACT therapist is attainable. There is no particular or special gift needed, but it will require an investment. As noted by Kottler and Carlson (2014), becoming a master at therapy is achievable. In ACT it requires not only grappling with the written material and delving deeply into the model, but also personal investment, living ACT in one's own life. Kottler and Carlson note the different qualities that define a master therapist. Each can be worked on, and I believe, is uniquely suited to the ACT model. For instance, they note that master therapists perceive the underlying structure of situations, noting the deeper issues that are creating the problem for the client in a faster and more efficient manner. As well, they have a unique way of persuading and influencing clients with remarkable effectiveness. That is, they are more rapidly and better able to make sense of meaningful patterns of behavior.

Carlton's Reflection

Is persuading at odds with allowing people to choose?

Robyn: *I think we always want to take the stance of personal choice in ACT. The client is the one who ultimately chooses what will happen with their feet. As a therapist, the act of persuading or influencing is not about power, but more about what's revealed in the etymological origins of the words: to persuade is to pleasantly urge, and influence comes from the Latin for "a flowing in." If a client is stuck or their behavior is ineffective regarding their values, then I hope to pleasantly urge them into the flow—back into life, into action—the client choosing what action but fully committing to taking the steps to bring values to life.*

ACT, being a behavioral therapy that looks to the function of behavior and incorporates how language influences and participates in human suffering, creates the context for describing "underlying" and "deeper" issues. Understanding human suffering, as it is linked to verbal behavior and avoidance of internal experience, is a fitting model. Recognizing our attachment to mind and how this attachment pulls us out of the moment, how it shrinks lives and pulls people away from what they care most about, sets the ACT therapist up to capably influence clients, especially given the therapist's reliance on client experience.

Two of the most important aspects of ACT involve (1) understanding the function of behavior and (2) setting the overarching context of ACT: being in motion, using your feet. I'll expand on these concepts below.

1. *Understanding the function of behavior.* Conceptualizing clinical cases from a functional standpoint (i.e., looking for patterns of behavior across time and context, including the interpersonal relationship and intrapersonal experience) sets the stage for detecting problematic behavior and intervening in rapid and effective ways, particularly as behavior change is linked to meaningful life outcomes. Behavior tied to its function can be quickly targeted for modification. Indeed, working with control as the problem and creative hopelessness in conjunction with values, for instance, has the power to potentially create behavior change at session one. Additionally, psychological flexibility, as practiced by the therapist, allows the therapist to tolerate a high degree of complexity in human behavior and ambiguity as experienced by the client and, at times, throughout therapy sessions. Undoubtedly, acceptance of both complexity and ambiguity are also part of the mastery experience.

2. *Setting the overarching context in ACT: being in motion, using your feet.* I have come to hold that the ultimate goal of ACT is to stay in motion—an aspiration for both you and your clients. Values cannot be lived from a "still" position. We are carried into and through our values with our actions. In other words, I am arguing that life is found in the feet. It is what you do with your feet that matters. It matters whether it is about putting one foot on the floor in the morning when you do not feel like it or whether it is about putting one foot in front of another through a painful divorce; a difficult social, family, or work situation; or the anguish of loss or existential struggle. All we have is motion or *being* in motion—open conscious living.

   We are the only species that will stop moving based on emotion or thought experience. This is not to say that other animals won't freeze under conditions of actual threat. Although humans might do the same, they are the only ones who will do it based on a *thought* of threat (rather than a real danger) or feeling of fear or anxiety, as far as I am aware. We can get stuck and begin to hold still in our literal and metaphorical efforts to control our experience. But what else will we or can we do?

Life has to be about motion, creating motion, and at times, any motion. It is the only thing that can break up a rigid and repetitive system that is stuck or holding still. As well, in ACT there is an invitation to do this movement with openness to experience and conscious awareness. Living life from the feet up is the metaphor for this process (and is part of the therapeutic stance); it is the assumption I hold and operate from, running through every thread of the tapestry of therapy. It is sewn into the fabric of an ACT approach, carried into therapy and worked on nearly from minute one.

Living life from the feet up means action. It says when you are lying there without a hope in the world or a sense of a way to face the day, you put your foot on the floor…and you step…and then you step again. Each step, each movement is then brought to life or characterized by meaning, by values. We are so used to living life from the head down, trying to think our way into meaning, that we fail to create it with the feet; it is almost as if we have forgotten our feet exist. This failure is where we get stuck, and this is where our clients get stuck. Meaning is in behaving. You and your clients are invited to live life from the feet up. Movement is fundamental to the work in ACT and is the soul and anchor of every session. The three pillars do not come to life without this provision. Taking action to practice openness and awareness is part of a fundamental framework supporting all of the processes. Movement is required. I assume that every client will need to engage in movement, and that motion is not in a changed thought or powerful insight; it is in changed action. Action emphasizes the third pillar—being engaged—and is the glue that bonds the pillars together.

Let's look closely at each of the three ACT pillars: open, aware, and engaged.

Willing and Defused: Practicing Openness by Staying Present with Ourselves

Becoming open involves the process of actively and consciously choosing to release oneself into full acceptance of the internal stream of present moments. In openness, the first pillar in ACT, we "make space" to experience emotion and mind as it unfolds in the here and now. Your personal experience of releasing yourself in this way, or "space making," will move you toward mastery and the heart of ACT. Making space may be likened to allowing wherein we contact—with no effort to resist, avoid, or escape experience, including both pain and joy—internal experience. We relinquish any attempt at excessive and misapplied control, allowing both mind and emotion to rise and fall as they come and go—the ongoing flow of experience and mind observed. Willingness and defusion are the avenues to openness in ACT.

> ### ACT Core Processes: Openness Pillar
>
> - *Willingness is a process of fostering acceptance while undermining the dominance of emotional control and avoidance.*
>
> - *Defusion is the process of undermining language-based processes that promote fusion with "mind," needless reason-giving, and unhelpful evaluation that cause private experiences to function as psychological barriers to life-enhancing activities. It is the process of observing ongoing thinking.*

Openness can be practiced in varied activities and through many avenues. Openness might involve focused meditation or actively saying yes or no to a challenging circumstance. It might involve doing something never done before or ceasing to do something always done. It, without a doubt, also involves the practice of nonattachment. As I share a couple of personal stories in the following sections, I invite you to consider your personal stories as you read. Note your reactions, paying particular attention to how your stories capture you across time. Notice, as well, the personal impact of being captured and how that may influence your behavior across contexts, including therapy.

In an all-day mindfulness retreat I attended, the teacher, a once-upon-a-time Buddhist monk, initiated the day by asking all participants to give up their hopes and dreams. Not merely to let go of the desires for what might come from the day, but to give up all hopes and dreams, every single one, for an entire life. This request struck me in a way that I had not encountered before. A disquiet set in that rapidly progressed first to an anxious sense of loss and then to deep sadness. I began to cry. At times, I sat with small tears and at times I was engulfed in giant sobs. I didn't want to give up my hopes and dreams. I cried throughout the day. I cried through the walking meditation. I cried through the qigong. I cried through lunch, and I cried through every single body scan and sitting meditation. The words "give up your hopes and dreams" rolled through my head over and again as the day moved on. "Give up your attachments; let go of your ideas about what could or should be." By the end of the day, I had tentatively touched an openness, a free space where hopes and dreams could be held lightly, moving in and then flitting away. A real sense of letting go and releasing into full acceptance of the internal stream of present moments arrived. Ultimately, this experience turned out to be one of the more reflective and insightful I have had, not just at a personal level, but also in recognizing what it is that we are asking clients to do when we are asking them to *let go*. We are asking them to give up their hopes and dreams. We are posing an alternative to control and an idealized conceptualization of themselves, their past, and their futures. Giving up hopes and dreams is a big ask. It is a big ask of our clients, and it is a big ask of you (yet, the ask remains and is there for you to do). But when done with compassion, inviting ourselves and our clients to open to emotion, to experience, to the ongoing flow of thought, can be liberating.

> ### Carlton's Reflection
>
> *Asking clients to give up their hopes and dreams is a fascinating idea and seems to correspond with a Buddhist perspective on how each of us fundamentally relates to our sense of self. It reminds me of what Pema Chödrön (2000) has written about this subject: "Hope and fear come from feeling that we lack something; they come from a sense of poverty. We can't simply relax with ourselves. We hold on to hope, and hope robs us of the present moment. We feel that someone else knows what is going on, but that something is missing in us, and therefore something is lacking in our world" (p. 50). This orients me directly to the Buddhist notion of nonattachment, which might be exactly where one needs to be to grasp (I just realized the irony of using that word!) what openness is. But how many people in therapy will (want to) go that far? Is this a departure point from therapy to spirituality/religion? Hope has a lot of currency in Western civilization; I'm not sure it's something that people are willing to give up, or perhaps are often in the right place to give up. But I think it is where one might ultimately need to be to* more *fully appreciate the meaning of openness.*
>
> Robyn: Not all clients will want to go that far. For some, this notion will be anxiety provoking or particularly scary. The therapist can invite the client to be open to that experience as well. I do always stand in hope for the human being. I have hope for the person in pain. I don't hold hope for their attachments. And yes, for some this will move into the realm of the spiritual or into religion. I hope that as therapists we will not be afraid to venture into these areas. We have far too long stepped away from these matters, dismissing them because they are not science. It is a mistake. Many clients are guided in values and living by their spiritual and religious experience. When we can explore these openly with the client, we are working from our nonattached place and providing the safety and presence for clients to bring these fundamental aspects of themselves into the room.

Of course, I am not saying, following this day-long retreat, that I somehow am now able to be open in this way at all times. Indeed, I am largely not. I get caught up in the day-to-day, the pains and struggles, just as anyone else. I routinely work to reorient myself to willingness and defusion. It is part of a practice in life. Not as two separate entities, first one and then the other, but as a whole (defused and willing simultaneously), as an open stance. Letting go in this way is there for our clients to do and it is there for us to do.

THE RISE AND FALL OF OPENNESS

Knowing that I will get caught reminds me of one of my favorite book titles: *After the Ecstasy, the Laundry* (Kornfield, 2001). It's a book about many things, but key to this point, it's about how enlightenment is followed by, well, life (and life is always in motion). In the same way, being open is followed by being closed. Being free is followed by being captured. We collapse back into our attachments. Holding to openness as a process is essential—observing the intertwined stream of emotion and thought is to be inexorably engaged over and again. There is no end to giving up on hopes and dreams.

Letting go of attachments can feel more or less significant as well. I will have times when I am entirely attached—unwilling and fused (inflexible). As I relate this next story of a smaller attachment, I invite you to consider your own smaller attachments. A colleague I used to work for was editing some of my writing, and I had received the revised document back multiple times with a large number of edits. When it came back yet another time with many corrections and deletions, I got hooked. I began to defend. *I have done a lot of writing*, I said to myself. "I have edited books and manuscripts and presentations," I remarked to my colleague. "I have been around the editing block, you know." Something was happening. It felt threatening. My abilities were being questioned. This bit of editing feedback grabbed me: What if each of the changes in the document was reflecting my abilities, my character, my wholeness? Captured.

Now to me, this was a smaller issue among issues, but I still had to work at it. I had to consciously let go. I practiced defusing from both the judgment of other and their edits of my work and the judgments of myself. Willingly holding the fear, opening to the seeming and felt threat, I let go by gently watching the thoughts and feelings rise and fall while accepting nearly every change in the document and thanking the editor. The material was better for the changes as it turned out anyway. My clinging was palpable, potent, but short. Still, I suffered inside of it. What smaller things do *you* find yourself suffering inside of? And how do these play out in your life? How do they show up in the presence of your clients? Are you aware of them and how they might influence the client on a regular basis? Letting go, forever, is for the small things too.

> ### Manuela's Reflection
>
> *What smaller things do I find myself suffering inside of? is a very evocative question for me. What are the small or subtle ways that getting captured creates suffering in my daily life? Recognizing these small attachments challenges me to explore how open, conscious, and engaged I am in general and in the moment-to-moment experiences of life. Bringing this question into my mindfulness breathing practice will help me recognize these places. It is not about the answer I get; it is about the perspective I open up to when asking.*

And what about the *biggies*? What about the ones that capture you, pull you in, and take you for a long and painful ride? What do we do when we get caught by our deepest fears? As a person who has spent the better part of her adult years developing her career, threats and worries about it can run deep. They can grab me in such a way as to keep me up at night. What are yours and do they catch you in such a way that they push you around? Get you to hold still when movement is required?

Being bullied as a child, from ages eight to eleven, created three incredibly painful years. I was called names (e.g., "four eyes," I wore glasses), I was forced to tie people's shoes, and I was told to "Go back to Mars where you came from." My eyes poked and stomach punched, I was tripped and pushed numerous times. At recess, I would be the first to the door so that I could run and hide behind the school buildings to escape the torture. I often went home with a "stomachache" (Mom always made me go back—and I am stronger for it).

That learning history was powerful. Today, when I feel forced or coerced, when I get that old feeling of being bullied, I can get powerfully hooked. I collapse into a sense of not being good enough, and I brace myself for the impending humiliation. However, when I am open and aware, willing to feel the feelings and practicing defusing as I show up to the thoughts that arrive when this learning history is triggered, I can either struggle or let it be as it is and take action that is closest to my values given the circumstances. I can observe these experiences and move my feet.

> ### Manuela's Reflection
>
> *I experience myself growing emotional as I read this story. It is a bold move to put it on paper. What you are speaking to is so fundamental to ACT. I can feel the risk; I feel an anxious sensation inside of me while reading the passage, but also the complete bodily feeling of it being just the right thing to do…hold and move. And what moves me about this story is seeing someone face suffering directly. It takes courage.*

> *In my own experience when starting ACT, learning to come face to face with my suffering was challenging. It was also challenging facing my clients' suffering. I could feel the urge to calm and comfort them and make them feel better. And over the years of practicing this, like the bold move you are modeling by sharing your experience, I am now more willing to be with my own and my clients' suffering without defenses, and I realize the freedom that comes with it.*

When I get hooked, however, I can find myself seeking reassurance. Now there is nothing inherently wrong with that unless of course it goes too far—unless the seeking of reassurance falls out of balance and, driven by fear, inspires me to do something linked to an apparent understanding of myself that is about not being acceptable at a fundamental level. As you might expect from an ACT perspective, it doesn't solve the problem, and indeed at times it can increase the experience of humiliation. After all, if I must seek reassurance, then I must be operating from a place where I am cast down; engaging in the reassurance behavior itself can be humiliating—the very thing I am trying to avoid. It is in times like these that the ACT work is hard. I need to practice, sincerely and with effort, bringing myself back over and again to openness, to feeling and defusing from thought.

Giving up my hopes and dreams here is complex. It may mean doing things that are potentially difficult or scary. It may mean taking a stand against someone who is trying to intimidate me, or it may mean getting out of the way when someone tries to force me to do something. It may mean accepting that others may not always like me, and it may mean sitting with the experience of humiliation and doing nothing about it. Feeling it, connecting to it, being aware of its qualities and the urges that come along with it: these are difficult experiences. They capture me. My work concerning them will be lifelong…as will yours and your clients'.

REFLECTIVE Practice 2.1

Jot down a couple of small attachments—places where you get hooked and start to question, judge, and defend yourself.

Questions to explore:

- How often do I get attached?
- What hopes and dreams are inside of this attachment?
- What is the impact? What happens?
- What do I say to myself? What do I say about others?
- How old are these attachments?

- Do I ever wonder why I am still attached?

Notice how they rise and fall and how they will rise and fall again.
Now consider a *biggie*. Write a few sentences about it.
Questions to explore:

- What hopes and dreams are inside of this attachment?
- What are my deepest fears here?
- How do I step away or turn away from myself when this arises?
- Do I expect it to resolve?

Notice how it will rise and fall, and it will rise and fall again, forever.

OPENNESS IS A WAY OF BEING

I talk about these personal experiences above not to make the point that you and your clients need to have a day-long retreat filled with tears to arrive at a true sense of letting go. But I might argue that finding your personal and ongoing process for releasing into full acceptance of the internal stream of present moments may be part of what is required, or at least contacted, to fully convey and understand what it means to be open for you and your client.

Carlton's Reflection

I wouldn't be too quick to dismiss the possible necessity of a day-long retreat filled with tears or a similarly powerful transformative experience, to fully contact what it means to let go of those things that capture us. Think of the extensive learning history that each of us has had. Years and years of being caught, again and again, and again. Indeed, we sometimes even define ourselves as the very things that catch us! If this is to change, then it's quite possible that something momentous has to happen. Exactly what that "something" is might differ from person to person; it might need to be a sudden insight, it might need to be a dawning realization, it might need to be something far more heartfelt, but I think the chances are stacked against it coming from a cursory experience of any kind. And then, even if someone has encountered such a transformative experience, there is indeed the "laundry"; there is the constant motion between capture and release, and the commitment required to let the experience continue to be a guide. As ACT therapists we look to create a transformative environment within the therapy room, but is this sufficient for the kind of change required to let go of such things as our hopes and our dreams, or does this need to be supplemented by something else?

> Robyn: Interesting. I think I have hesitated to recommend such an endeavor as I want to convey that there are many ways to practice "letting go." And indeed letting go can be a powerful experience. But I also believe it can be a quite small experience. Darrah Westrup reminded me of this in our book, The Mindful Couple (Walser & Westrup, 2009), when she wrote about letting go of whether the toilet paper rolled from the top or the bottom—a longstanding disagreement she had with her husband. But the big stuff? Not so easy, and something much more robust is likely needed in those moments. ACT therapists who are looking to master the therapy may want to consider the venture seriously (and be reminded that process is the outcome).

An honest look at where you collapse into your mind and get swept away by emotional urges is part of what is needed. What are some of the stickier places, both small and large, that get you stuck? Take a moment to reflect on the ones you considered while reading, noticing not just the painful experience or set of events that lead to this stickiness, but also what you do in response to them (consider Reflective Practice 2.1). I invite you to ask yourself these questions as you consider these sticky places: Have you ever thought, *I thought I was over this. Why is this still catching me? Why can't I just let this go? Why is this bothering me so much? I know how to observe my thinking; why can't I move on?* Some who read these questions might say, *Yes, I know these places and know that I continue to get stuck.* Others might read these questions and wonder why I am asking them at all, given the work we do inside of ACT. They are obvious, maybe even routine. If you find yourself here, even just a little bit, then dig deeper and let me ask further, why do you still struggle? Why do you get angry? Why do your buttons get pushed? Why do you have difficulties in relationships? Why do you, at times, find yourself reacting to feelings of anxiety, fear, and loneliness? It is in these places that we sit in our humanness, wanting to be free, wanting to be enlightened—wanting to be open—accepting and defused. Let me repeat the words of my one-time-Buddhist-monk teacher: give up your hopes and dreams.

Working on openness will never provide a unique or magical release into experiencing the ongoing present moment that isn't followed by being captured—followed by doing the laundry. There will not be a time when you or your clients will be free from future struggle. This awareness can hinder or help ACT therapists. When considering letting go of hopes and dreams, be cautious about being drawn into hopelessness, a place where nothing is there to be done. Something is always there to do. Therapists have noted to me that telling their clients to give up their hopes and dreams, literally or metaphorically, seems like a cruel request. I might posit that it is the opposite. This request is about nonattachment; it is about being defused and willing. It is from this free place that we can settle into being whole in the moment. If there is no need to have something else, to be something else, or to be somewhere else, if you do not need to feel or think other than what you are feeling and thinking in the moment, then you are undivided, you are as one—you are whole.

It is from this place of wholeness that in-session work with the client can be transformative. It is liberating. It provides the "place," being in our skin with our history in every context where we can actively and consciously choose to release into full acceptance the internal stream of experience.

> ## Manuela's Reflection
>
> *It is truly transformative to find the value in nonattachment, but it is also not a place where you arrive. Correct? For me, this is more like catching a glimpse of something. And, even from inside these very short glimpses, I can take a stance of wholeness. Is it the same with clients? I believe yes. It is a place to point to as therapists. It is like pointing to the breath in a breathing meditation. And when you get distracted, you gently point again. Let's take a look at this using my experience with one client. He came to therapy at the age of sixty-two feeling very damaged due to multiple severe traumas in his life and a long story of depression. His psychiatrist, with whom I talked after the first session, told me that the client was a chronic and severe "case" and there was not much left to do. Both client and psychiatrist had collapsed into their view of "damaged goods." And during the first sessions of working together, I was able to notice how easy it was for me to also start seeing him as broken or as a person with something missing or as a problem to be solved. I can still get stuck there from time to time. And when I found myself not being able to "solve" him, I became the damaged goods, a bad or broken therapist. Supervision was the way for me to gain flexibility and to be able to see the client and myself as whole. It was not easy. My client and I were doing the best we could with our learning histories. However, we both needed to drop our hopes and dreams of being something different from what we were, and when we did that from a stance of being whole, change became possible.*

As an ACT therapist, the request to continually give up your hopes and dreams is a way of being. It is a chosen practice that will assist you in your work with clients. It will ground you in your wholeness, creating a way to work with clients from an experiencing (meaning you are open to your experience) approach. It will help you to hold the judgments and thoughts, the mind chatter about your therapy and the emotional reactions to your clients, lightly, freeing you to be bold in your work: to compassionately confront, to say what you have been afraid to say, to work honestly and with integrity in assisting your client to move. The goal is not only to help your client be flexible but to sit in openness, working with willingness and defusion processes to assist the client in moving from being closed and unwilling to open and free. It will allow you to be fully present while focusing on another. It is a choice as well as a process.

Conscious Being: Personal Practice of Awareness

Awareness is waking up to this moment, contacting I/here/now from a sense of self that is consistently present and able to observe what is sensed, felt, experienced; this is the second pillar in ACT: aware. It is a sense of self wherein the world encountered is only possible from the world of "you." I am here in this moment experiencing, and I am aware of my experience. Wherever you go, whatever you do, whatever perspective you take, there is a consciousness that goes with you. You are the "middle," the center of your experience. It is with conscious awareness that we can observe our experience; it is from this center that we show up in each moment defused and accepting. *Consciousness* to flow of mind and emotion intimately links us to freedom, choice, and nonattachment.

ACT Core Processes: Awareness Pillar

- Contact with the Present Moment *is actively working to live in the here and now, being aware of the present moment, contacting more fully the ongoing flow of experience as it occurs.*

- Self-as-Context *is the process whereby the individual makes contact with a deeper sense of self that can serve as the context for experiencing ongoing thoughts and feelings. This kind of perspective taking may be key to flexibility, compassion, empathy, and other qualities of well-being. Self-as-context is distinct from the content of an individual's life. Not being defined by their thoughts, feelings, and sensations, a person is a vessel in which these experiences occur.*

PAVING THE WAY FOR FLEXIBILITY

I would submit that awareness to experience is the very thing that invites flexibility. The practice of awareness takes us beyond the limitations of thought. As well, awareness has a soft but steady quality and is characterized by a kind and quiet release into I/here/now. It is not a firm and rigid practice of attending. Chödrön (1991) describes a connection to awareness that involves not only a gentleness but also an allowing that is awake to experiencing "how large and fluid and full of color and energy our world is" (p. 27). Touching awareness takes practice. We easily fuse with thought. And as noted earlier, routine practice is part of mastering the work done in ACT.

If a masterful therapist is one who allows, without any prohibition or hindrance, the presence of the complexity and ambiguity found in therapy, then it seems that conscious awareness of these experiences would be required. The layers of experience and varied patterns of behavior seen in a single session or sessions point to the very quality of complexity. I want to add that this complexity and ambiguity exist for you, the therapist, as well. The emotions, thoughts, and sensations—the intrapersonal process—that you experience in response to the interpersonal process (explored in chapters 7 through 9) in the therapy room are vast and borne out of your history interacting with another's history. If you are unaware of these processes, if you are fused, your ability to be flexible has been lost. You may find yourself narrowed in on the wrong issue, chasing client stories rather than the function of their behavior. Or you might find yourself responding to a client in ways that are about your pain rather than an effective intervention for the client. You might also find yourself lost in thought; you are no longer listening. You might see yourself caught in an earlier experience from the day; you are no longer in the room. Awareness to your own experience and ability to bring yourself back over and again in the moment matters.

LISTENING BENEATH THE WORDS

Consciousness to your internal experience (intrapersonal process) and consciousness to the experience between you and your client (interpersonal process) is key to "listening beneath the words," detecting patterns of behavior and responding in more efficient and effective ways that speak to the function of a behavior rather than merely its form. I might also submit, at this point, that the distinction between inner experience (intra) and outer experience (inter) is, in one sense, artificial. The two are intertwined; this is part of the complexity that we are called to observe, notice, accept, and be responsive to. Choosing what experiences and information to act on in any moment depends on this awareness and creates a more flexible and larger capacity to respond to the function of a behavior.

Again, I pose some questions for your consideration. Do you routinely practice some form, be it formal or informal, of returning to the conscious awareness of experience? Do you meditate or engage in awareness of movement or awareness of the body (e.g., yoga, qigong, tai chi)? Do you practice any kind of pause or effort to slow down just enough to move into awareness? To soften into what you, as a being, experience? I hope that the answer is yes. If not, I make a gentle invitation to explore what prevents you from stepping into this work. I have been given many reasons across the years, with the most significant and challenging one being "I don't want to." Others include "It's too hard," "I don't have time," and "I don't think it is needed." I will be the first one to say that it is up to you. But I will also be the first one to say that not engaging in some form of routine practice, formal or informal, that brings you to conscious awareness will also limit your capacity to develop mastery. Regular existence outside of awareness, with no effort to observe, limits flexibility. Options and possibilities are closed off or unavailable. Your ability to respond to a particular context in a particular moment is lost to your being captured by mind. It is already hard enough. We spend most of our day fused, living inside the mind, ignoring the body and its sensations, rarely contacting the flow of thought and experience. You don't have to look too far to see how this works. A few simple questions will suffice: Where did you spend your time today? Was it mostly in your head?

REFLECTIVE Practice 2.2

Consider your personal formal and informal mindful awareness practices and ask yourself these questions:

- How often do I practice?
- If I have not been practicing formal mindful awareness, what gets in the way? Is there a cost?
- Do I invite my clients to practice mindful awareness?
- Do I practice mindful awareness during a session, not as an exercise, but as a process of therapy? If yes, how does it help? What is its effect?
- If I do not practice, what are my thoughts about asking my clients to do it anyway?
- Is there a commitment that I might make about my practice? Write it down.

Any time we spend the majority, if not all, of our time in our head, our flexibility decreases. The broader context of experience disappears. When we are not aware of our experience, our body, and the relationship between us and the "outer" world, even the simplest of choices can disappear. I didn't realize I was hungry; I didn't hear my friend calling my name; I ate yet another doughnut; I was late for my meeting; I didn't hear the bird outside my window; I didn't see the sun splash on the greenery in the backyard; I didn't feel my dog lying next to me; I didn't recognize my tiredness; I didn't taste my meal; I didn't see my "ego" get caught in that struggle; I didn't connect to my sadness, my anger, my disappointment, my joy, my love, my surprise. I spent the day in my head and now the day is gone. We not only need to help our clients to "hear the world beneath the mind," we need to listen ourselves.

And this next piece seems quite important. If we are encouraging others to live by their values, I assert that awareness is required. Being aware allows us to gently observe and choose outside of the fused experience that whittles at our day and pulls us away from the things we care about most. When we are unaware, we are no longer observing the ongoing flow; rather, we are caught up in it. It robs us of our choices and the capacity to be with ourselves and others fully and presently.

> ## Manuela's Reflection
>
> *Reflecting on our awareness in this way is so beautifully expressed. I can feel myself also wanting to move more fully into listening beneath the words. Awareness is a tricky place. It means more than concentration or not being distracted. In my clinical work, it means catching myself when I get in autopilot mode or become ego driven. And with infinite patience for myself, come back to awareness, come back to the moment. I have also noticed that working on awareness helps me not be caught in the client narrative or my own. I can switch to functional analysis more readily. For me, developing awareness has been a prerequisite to being able to focus on functional analysis and the flow of therapy. It also allows me to profoundly connect with the client's suffering.*

Awareness of our experience also allows us to connect to and have empathy and compassion for the experience of others (as well as our own). It is in this place that our therapy is alive. When we engage in our practice of awareness, we create a platform for intense listening, allowing us to hear not just the words, but the smallest tremor in the voice or change in its intensity. It allows us to see the tiniest of shifts in facial expression or the slightest of adjustments in the chair. It will enable us to remember the stories and experiences of the client from one session to the next in greater detail, noticing a change in appearance, informing our assessment of behavioral patterns, informing our understanding of the client and her struggle. As Julian Treasure (2017) notes, conscious listening allows us to live fully connected in space and time to the physical world around us and connects us in understanding to each other.

Listening with conscious and deliberate awareness in this sense involves hearing the words, the emotions, the body language, and the message beyond the words. It allows us to connect to that which is subtle, quiet, or understated. From this place, our capacity to offer empathy and compassion grows. When we place a premium on being here and now with the client, we better understand their experience. As we connect to our own larger sense of self, we see a sense of them, a wholeness, that is beyond their emotional struggle. We see a sense of them that is beyond a single story or our labels for them. We better connect to the ongoing flow of their experience, placing us in contact with the larger sense of self that is them—their "beingness." As we are aware of our own, we are aware of their consciousness. It is here that you and your client can stay, holding pain, together.

> ### Carlton's Reflection
>
> *Reading about self-awareness is like a thundering wake-up call. I realize that with some of the clients I have worked with, my self-awareness has waned at times, my full awareness of the client compromised. I think there is so much in the context bearing down on us as therapists that can push us into sleepwalking through therapy sessions. We get caught up in our thinking; we form perceptions, assumptions, and beliefs about the person opposite us; and although these can be helpful at times, they can also pull us away from noticing detailed aspects of the client's behavior (in the broadest sense of the word). These moments of noticing, when kept to ourselves, can guide our responses, and when made explicit, can sometimes function as important turning points in therapy. Sometimes the awareness of a slight change in facial expression can be more important than the hundred words that preceded it.*

> ### Manuela's Reflection
>
> *I love this connection to beingness. It reminds me of a quote by Carl Rogers (n.d.):*
>
>> *When I can relax, and be close to the transcendental core of me, then I may behave in strange and impulsive ways in the relationship, ways I cannot justify rationally, which have nothing to do with my thought processes. But these strange behaviors turn out to be right in some odd way. At these moments it seems that my inner spirit has reached out and touched the inner spirit of the other. Our relationship transcends itself and has become something larger.*
>
> *As Rogers conveys, we are no longer what we say we are, nor what we should be. We are no longer what our thought processes say. Recognizing this can help the client and me dwell in a more experiential space, learning from it by experiencing the direct contingencies. Increasing both our abilities to discriminate the function of our behaviors. We will be in a better place concerning predicting and controlling behavior and flexibility.*

Life from the Feet Up: Personal Engagement

The third pillar of ACT is engagement. Earlier we focused on how engagement, in a broader sense of the word, binds the three pillars of ACT together. In this section we focus more specifically on the pillar itself. I should start by noting that being busy is not necessarily being engaged. You and your clients may be doing many different things throughout the day—taking the kids to events, completing projects at work, traveling, exercising, running to the store, visiting with a friend, hanging out with a partner, and so on—seemingly engaged in all of them. But again, being active does not equal being engaged. Engagement is deeper than daily activities that fill up our lives. It's deeper than a single goal or set of goals. It is about orienting our lives to a set of larger aspirations linked to creating personal meaning. And meaning is directly generated by what we do with our feet and their purposeful intention—yearly, weekly, daily, and hourly. Our sense of purpose is being created across time and found in what we do in each moment.

ACT Core Processes: Engagement Pillar

- Values *involve identifying valued processes in living, such as being loving, being intimate, connecting or belonging with others, kindness, generosity, honesty, integrity, growth, humility, and discovery. Numerous values lived inside of important categories such as family, friends, romance, spirituality, community, health, and environment bring vitality. Each is a guide for creating meaningful life directions.*

- Committed Action *takes you there. This process involves building larger and larger patterns of behavior that reflect chosen values. Note that there is no arrival concerning values. Individual actions in the service of values can always be completed, but there is no end to engagement in the meaning created when you act on your values.*

If it is in conscious awareness that flexibility and freedom exist, I then ask what do we do with this freedom? One of the most personally influential books I have read in my life is *Man's Search for Meaning* by Victor Frankl (2006). He writes: "…we had to teach the despairing men, that *it did not really matter what we expected from life, but rather what life expected from us*" (p. 77). We create our meaning. In our freedom, we can choose to bind ourselves to what we value—a kind of promise made to honor, by our actions, what we hold most dear. And I would also submit that this be done with a sort of passion and intensity— an engagement. Not in the sense of crazy, unbridled enthusiasm for all that you do, but with the understanding of what is meant by commitment with intention: an act of binding yourself to a course of action, over and again. No matter what challenge is faced or struggle is presented, your feet are pointed in the direction of values and movement is involved.

I hold this to be important not only in our personal lives, but also in therapy and as a therapist. Your values and the actions taken to bring them to life count out there in the world and inside the therapy room. Owning your motives in therapy is potentially the truest path regarding your relationship with the client. You may have been told to keep your values out of the therapy room. I would argue that therapy is not values free. Indeed, I worry when we assume it is. Helping others, itself, is based on a value. I am aware that I want something for my clients, and I make them aware of that want. I hold fast to choice—I want them to have freedom. I hold fast to capacity—*something is there to be done*. I hold fast to ability—they can respond and be responsive. It is essential to know what matters most to your clients, but it is also important to understand what matters most to you. As you take action on your values in session, you model for your client. Being compassionate, loving, and kind are all there to be engaged.

Your values will play out in session in other ways as well. For instance, if you hold authenticity and honesty as personal values, then these will assist you in orienting toward the "truth," for yourself and the client, even in the face of great emotional difficulty. Engaging these values, while remaining open and aware, may give you the courage (as an action) to speak, confronting deception and other challenges—yours and the client's—that may emerge in therapy. Values-based action is part of being bold in therapy and guides the risks that you might take as a therapist. A risk guided by values is less of a threat than one governed by none. Holding values in therapy also allows you to have an ethical compass. And your awareness of these values will guide you in nuanced ethical matters, stopping you, slowing you down, leading you to seek guidance, support, or other kinds of direction when you have an ethical concern. If you have humility as a value, then you will be grounded in therapy, willing to bend, kneel, or bow as needed—creating a space where you and the client are the same, both human, both having your measure of suffering. Indeed, Kottler and Carlson (2014) note that humility is a crucial characteristic of extraordinary professionals—perhaps a piece of mastery and heart as well.

I suppose it goes without saying (so it's odd to now say!) that foisting personal values on the client is problematic. Honoring their values is essential. Extracting your values from the intrapersonal and interpersonal space that you and the client occupy is nearly impossible. Instead, they can be brought to life, engaged between you, while you fully recognize your and your client's personal capacity to choose. Flexibility is also found in the awareness, between you and the client, of what matters. You can each engage respectfully in the interplay of discovering not what matters in life, but what in life matters to you and your client.

It is essential to keep in mind that engagement can fall out of the realm of process when you believe you have arrived or when you think activity itself equals engagement. It can happen to us personally, and if we're not diligent, it can become the focus of therapy. At the beginning of the chapter, I noted that life is about movement. But not just any movement. It is movement linked to mattering. It is movement connected to aspiration. If we have only this one life and if it is short and possibly meaningless, what meaning will you create?

STAYING CONNECTED TO MEANING IN THERAPY

When working with clients, staying connected to meaning is key to changing the context, situation, or behavior. One of the most useful ways to get unstuck, to move forward, or to understand what might be happening between you and the client involves linking the process in therapy to what they would like to do with their life. Making this link is tricky at times because you and the client might have different views about the goal of therapy. Therapists may find themselves struggling when the client's agenda is not lined up with theirs. For instance, a therapist can be steadily working on a values-based process and find that the therapy is not moving anywhere. This lack of movement may seem odd given that it is generally the client's values that the therapist is working to bring to life. However, I have spoken to and worked with many therapists who have not quite caught, as it can be subtle, that the client is still working on a control agenda, or a different agenda entirely. One of the challenges regarding this issue is that it can happen reasonably deep into the therapy and after the ACT therapist has already been through work related to *creative hopelessness* and *control as the problem*. If the therapy is stale or your client seems checked out, it's useful to check and see if you are still working on the same agenda, the same desire to bring engagement with meaning (now and before thoughts and feelings change) into the client's life.

SOMETHING CAN BE DONE: COMMITTED ACTION

The key purpose of vital living and our short time here on Earth is purposeful engagement. No matter the dilemma, the struggle, the thought, the pain, something is there to do. From the smallest movement to the most significant change, some action can be taken to shift your feet, to shift the client's feet in the direction of meaning. There is no limit to the kinds of behavioral goals a therapist can work on with a client. These small changes can be supported and can happen right in session or be encouraged in multiple contexts outside of the therapy room. Clients will sometimes say, "Nothing can be done" or "I can't." I often reflect on the purpose of therapy if this is actually the case. There is no movement if there is nothing there to do. Working to support clients in their efforts to take a step, even if barely noticeable, is a change; it is a turning of the feet.

This work is there for us as therapists to do too. Whatever struggle you find yourself in, whatever dilemma or pain, something can be done. Will you turn your feet? The invitation forever stands: Be bold. Live life from the feet up.

REFLECTIVE Practice 2.3

Consider doing this Reflective Practice with a colleague or in supervision. Complete the practice and then openly discuss what you observed and learned. You may also complete this practice on your own.

Your personal open, aware, and engaged stance matters. Take several quiet moments to reflect on your "ego" (as it shows up both in life and in your therapeutic work). Explore some of the following questions, or ask your own. Notice how ego may play a role in each:

- Why do I do what I do?
- Are there things that I do for recognition or success?
- Are there things that I do to martyr myself?
- Are there things that I do to express dissatisfaction that are not verbal or direct?
- Are there things that I say to paint myself in a better light?
- Are there things that I do or say to get attention?
- Are there things that I do to look clever or smart?
- Are there things that I do to get away with something?
- Are there things that I do that I don't tell others about?
- Are there things that I do when no one is looking?
- Are there things that I do that are out of anger or unlike me?
- Are there things that I do that are about being included or excluding others?

Simply explore. Tap into an honest place. See if what you find there is where you want to change. Consider the time spent on ego issues. How can you use the practice of being open, aware, and engaged to assist you in your life process with respect to your ego? How can it be helpful in your therapy work to honestly explore these questions? What might you do with your feet when your ego has the greatest hold on you in therapy or in your personal life? Finally, how can personal openness, awareness, and engagement feed personal presence in therapy?

The Heart of Now and the Door to Next

Living life from the feet up is fully possible for you and your clients using the three pillars of ACT: openness, awareness, and engagement. You and your clients are invited to live life from the feet up rather than the head down, bringing your heart along. Personally engaging the three pillars in ACT sets the context for shifting your therapy to masterful intervention. Being aware of your experience builds compassion and caring as well as the capacity to listen genuinely. These are qualities of a masterful therapist working on the heart of ACT, weaving the processes inside the pillars into a fluid mix during therapy, nimbly transitioning between each. This is created inside of a personal process of engagement, an intent carried out through ongoing movement—practicing the model in your personal life.

It is with conscious awareness that we can observe our and others' experience; it is from this center that we show up in each moment of therapy and life. Awareness to flow of mind and emotion are intimately linked to freedom, choice, and nonattachment, for you and your client. Practicing conscious awareness is indeed a process—we have all found ourselves wrapped up in behaviors that we would rather not be doing. And we will continue to see that across time. The question isn't whether everything we do will become aspirational; the question is whether we can be aware enough to ask the questions of ourselves: Am I open? Am I aware? Am I engaged? And then point our feet in a different direction if we don't like the answer. And indeed, pointing our feet is about both the little things that we do each day and the big things that we do in our lives.

In chapter 3 we turn more fully to openness and awareness by visiting the three senses of self defined in ACT: self-as-content, self-as-process, and self-as-context. We explore these in a process of understanding and growing self-knowledge. If the practice of awareness is key to flexibility, then recognizing and being open to the experiences that have shaped our values and the ways we behave, respond, and relate to others will be part of the context of therapy. Knowing ourselves will be part of setting the context for change.

CHAPTER 3

Open and Aware: Knowledge of Self in Relation to Other

He who knows others is wise; he who knows himself is enlightened.

—Lao Tzu

When reflecting on the work I have done during therapy sessions, I often ask myself about the client and the possible reasons for their behaviors. In this reflection, it is valuable to consider the client's behavior in a particular context—the therapeutic session. Given that I am part of that context, my behaviors and their influence on the client are also worthy of consideration. How do my stance, my emotions, my sensations, my thoughts, influence what I do? How does my experience manifested impact the client, the session, the context? And, if context is defined as "…the changeable stream of events that can exert an organizing influence on behavior…includ[ing] both history and situations as they relate to behavior" (Hayes et al., 2012; p. 33), then perhaps I have my work cut out for me in answering these questions. Nonetheless, the work related to self-awareness or self-knowledge is imminently relevant to me as a therapist, to each of us as therapists. The experiences that have helped to shape us into the individuals that we have become today, the beliefs that we have developed, and the values that we hold are part of the context we participate in when sitting in the therapist chair. Knowing these experiences and knowing ourselves (to the extent that we can) is essential to modeling willingness, taking risks in therapy related to values, and setting the context for change. In chapter 2, I invited you to explore how your personal practice of being open, aware, and engaged—your own "movement" or process—contributes to mastery and connecting to the heart of ACT. In this chapter we continue this exploration by looking more closely at self-knowledge and the three senses of self described in ACT (see Hayes, 1994; Hayes et al., 2012) to assist you in considering your inner experience and recognizing the role it plays in your life and therapeutic work.

Manuela's Reflection

We are part of the therapeutic context. Not only for what we explicitly do in therapy and for how we relate to clients, but also in how we relate to ourselves. Our relationship with ourselves, then, is part of the context. The relating behavior we have with ourselves may influence how we work. Consider how we conceptualize ourselves and how much this conditions our therapeutic behavior. What aspects of ourselves do we hold too tightly to? What don't we want others to see? It's important to work with the answers to these questions in such a way that we are not driven by them. To be sure, in considering the relationship I have with myself, I have, from time to time, contemplated the question "Who am I?" It still gives me an uncomfortable feeling to ask this question given what I now know about ACT (i.e., any answer is only a conceptualization). However, I still ask. It helps me to be aware of my concepts of self and to work from a more embodied and contextual experience. I am more than any answer to the question "Who am I?" When I further wonder about who I am, I find myself asking, "Am I something?" or more specifically, "Am I 'some thing'?" It is here that I feel an even greater discomfort, a collapsing feeling in my body if we are truly some thing as conceptualized. Inside of this place, I have the experience that I must put myself in a cage. It reminds me of an exercise I did in an ACT course about completing the sentence "I am _____." I realized that I didn't have anything to say apart from describing some behaviors that happen in a context. Not being able to answer the question with anything but a set of behaviors, however, doesn't mean I have the sensation that I am not myself; rather, I don't completely identify with anything but the feeling of a perspective. This perspective is a felt sense, a place of witnessing, a consciousness. The other responses I can give of "some thing" are more like clothes, but I hold firm, in this consciousness, that I'm more than the cloth. Perhaps being aware of this seems like something a monk would do, but I don't think you have to be a monk to feel and live from this place. You can connect to it throughout your lifetime. Indeed, your clients can too. I have had the experience of working with clients who touch this experience, this perspective, without formal mindfulness practice. If you understand and help others to contact the verbal cues around hierarchical and deictic frames, you can create a place for this experience. Of course, I think that mindfulness practices are another way of touching this place and are an important part of practicing awareness in the moment.

Pursuing Self-Knowledge

One of my favorite ACT training activities is to explore the stories individuals hold about who they are. Asking individuals, "Why are you the way you are?" brings a plethora of interesting and varied stories ranging from "Mom made me this way" to genetics and evolution to childhood trauma, past experiences, and important adult relationships. Each story told with conviction has a rich and charactered quality. The surprise that follows when individuals are then asked to remember the stories that occurred on the fifth day after their eleventh birthday and how these events contributed to who they have become takes on a curious and often humorous quality—no one can recall those stories. Indeed, it is difficult, if not impossible, to remember the eighth day after your ninth birthday and the seventh day after your thirteenth. However, the stories of our lives were being written on those days too. We were being shaped and influenced by our environment; we were learning and growing on those days as well. Extensive experiential and verbal learning occur across time and context. This means that we are ultimately unable to entirely explain why we do the things that we do or why we are the way we are—we can never *fully* know ourselves through verbal explanations and conceptualizations, nor can we know the vast number of historical variables that shaped our behavior. Still, we can pursue self-awareness. It is paramount to understanding ourselves—our behavior, emotions, and thoughts—as thoroughly as possible. Gaining this kind of knowledge can be useful for many reasons: not only its impact on interpersonal relationships in the arenas of romance, friendship, and work, but also for the work in this book, in encouraging you to recognize how your behavior might impact the therapeutic relationship. The use of self-knowledge, including self-disclosure, emotional reaction in session and across time, and interpersonal processes, can be a powerful tool for change when used thoughtfully and when connected to function.

Carlton's Reflection

Fully knowing oneself can be an interesting therapeutic dilemma. On the one hand, clients often want to understand and have insight. They want to know themselves fully and seek therapist assistance in that endeavor. On the other hand, they often report that they know *why they did something.*

Robyn: *I agree that you can know some of your motivations for why you do what you do. Any behavior can be assessed for the consequences that sustain it in a particular context. But to know all the variables that have shaped your behavior is nearly an impossible task. I can never fully know your history. I can look at patterns of behavior in (the therapeutic) context to see what sustains them and ask, How do*

> these behaviors function? *and* What might I do to influence or shape them differently according to desired outcomes? *Nonetheless, the larger clinical struggle occurs when clients are too attached to the stories or reasons that they give as the causes for their current behavior (e.g., "I do what I do because I had a difficult childhood"). Stating that you can never fully know yourself isn't about disputing the facts of your life (e.g., you grew up financially poor, you were in a car accident at age fifteen, you have a specific genetic code, your culture was working class). Those facts can remain as facts. However, getting in a car accident at age fifteen is only one fact among millions. How did those—the millions of others—play a role? If I say to you [Carlton], for instance, that you make statements like "Clients want insight" because you are curious or being rebellious or simply like making these kinds of statements, it will be too limiting. You are much more complex than these labels or categories of behavior. If these assessments are "true," they are still only aspects of your experience. You are a complex being. To know each of the contextual variables, to fully know the learning history that shaped your behavior, is more than challenging.*

Even so, though therapist self-awareness is sparsely studied, findings regarding the impact of therapist self-awareness in session are mixed (Nutt Williams, 2008). On the one hand, research findings support a negative impact, with therapist self-awareness increasing anxiety and poor performance. On the other hand, therapist self-awareness has been associated with positive interpersonal experience and client perceptions of therapy (Nutt Williams, 2008) as well as favorable client ratings of the therapy process (Nutt Williams & Fauth, 2005) and increased helpfulness (Fauth & Nutt Williams, 2005). You may speculate, in the case of poor performance, that self-awareness of one's anxiety followed by judgment, as well as the desire to not be anxious as the therapist, might well create more fear and poor performance. If a therapist is new to ACT, these experiences may be further compounded, perhaps leading to problematic therapist behaviors employed to avoid the feelings of anxiousness. Possible examples are moving into a one-up position to regain "status" and control or quickly resorting to a technique, whether it is contextually appropriate or not, or imitating an admired ACT therapist only to sound inauthentic. It is here that self-awareness seems most important. Recognizing and opening up to one's anxieties, fears, and judgments; coming to know your vulnerabilities and strengths; and choosing behaviors in therapy that are in line with compassion, authenticity, and the discomfort of learning something new are all part of the process of mastering ACT.

Other thought-provoking data linking countertransference to the pursuit of self-knowledge exists. Therapist emotional reactions to the client, if not managed, can negatively affect psychotherapy (Hayes, Gelso, & Hummel, 2011). Knowing when to respond to your own emotional content elicited by the client is vital. Understanding when to use emotion and other kinds of self-disclosure in the therapy session can only be preceded by an awareness of these experiences and your relationship with them. Once aware, you can ask yourself about what seems to be eliciting the emotional reaction and if sharing this reaction in therapy serves a function that is useful to the client.

For instance, you may experience frustration or maybe even anger in a therapy session. Do you speak to this experience with the client? It depends. Being alert to the function of this experience will be important. I once worked with a client wherein I was experiencing a fair bit of frustration in the session. She sat across from me with a very noticeable and quite large bruise on her eye; it was black and yellow with fresh red marks. Her partner had struck her the evening before our session. I experienced all kinds of emotion, the most palpable being anger. I was angry at her partner, and my internal response was intense. I could feel my face turn red hot; I was sweating and tense. Should I use this emotional response in the session? I waited. I sat with the experience (although surely the client could detect that something was going on for me given my visible reactions). Awareness of my history and the in-the-moment experience was critical. Domestic violence was a part of my childhood background. In the session, in that moment, it was essential to be intentional and slow. It wasn't a time for me to put my feelings in the room about domestic violence. And there is something else that is harder to speak to if I'm honest with myself. I was also frustrated with the client. She didn't want to take any action. She wanted to return home and forget about it. I believe putting my frustration and anger in the room at this time would have only served to make things worse. Another angry person in the client's life was not what was needed. I breathed into my experience and returned my attention to the client. Compassion for her pain and stuckness seemed the best place to go.

I also once had a client who was stubbornly stuck on the idea that mindfulness should create a sense of peace and calm that should "never" be interrupted by anxiety if well practiced, and he buoyed his argument with a sense of superiority about this interpretation. After a couple of sessions of looking at this from all angles and perspectives as well as observing how stubbornness was ineffective in his interpersonal relationships, I put my frustration with his attachment squarely in the room. We worked directly on his wanting to be right and its impact on relationships with other. Many people in his life experienced a similar frustration with him. Sharing my frustration was a part of helping him to understand his interpersonal struggles. Letting him know his impact on me not only brought awareness to this problem, but also allowed us to work on broader issues—functioning in relationships. His desire to use mindfulness to escape anxiety was linked to being able to be connected interpersonally without anxiety. However, anxiety wasn't the problem; attachment to being right was.

These are only two examples, but they are legion. In the therapy room, every way to elicit emotional experience, the client's as well as yours, would be difficult even to imagine or write about here. The critical message is awareness to these experiences. You are an emotional being; to intimate that your emotions are not part of the therapeutic context is to miss your humanness. If you are attuned to the emotions that show up in the therapy room, you can effectively use them to assist the client and the relationship between you and the client. Willingness to experience whatever arises is part of this process. I can feel anger and not act on it when it doesn't serve the client. I can feel frustration and act on it when it does. Being aware of these experiences and knowing when to act, or not, is part of the personal development process. Staying connected to the function of using personal emotion in the room (in the service of the client) should be your guide.

I should note that there are times when the function is unknown, and you might be testing hypotheses. Here, you can still place your emotional experience in the room, but from a much more curious and questioning stance (e.g., "I am noticing a sense of frustration; I am unsure about why it is happening. Do you feel any frustration right now? Or is there anything frustrating about what is going on? Do others note frustration to you when you are caught up like this?")

REFLECTIVE Practice 3.1

You can do this exercise alone, but I recommend doing it in a supervision process or with a colleague.

Part 1

Consider a time or times when you have had a strong emotional reaction in session. See if you can pick a few specific instances. Explore the following questions:

- What elicited the emotional reaction?
- Did the client remind me of a historical event from my own life when it happened?
- Did I share this reaction with the client? If so, why did I share it? How was it intended to serve the client?
- If I didn't share it, why not? What might have happened had I shared it? If it would have served the client, what stopped me?

Part 2

Explore the following either by yourself or with a colleague(s): In your consultation with other therapists about your clients or your own supervision, do you explore your emotional reactions to the client(s) being discussed? Or do you focus on merely implementing the core processes? Consider bringing more exploration of your emotional responses from therapy sessions to your consultation work or into your supervision. Explore how doing so can inform your work. Process this with your colleagues, noticing and delineating the function of your behavior in relation to your clients and vice versa.

The Three Selves in ACT

The theoretical underpinnings of ACT can assist with growth in the self-knowledge process. Working to be open, aware, and engaged are part of this experience, but reflecting on the three senses of self (see Hayes, 1994; Hayes et al., 2012) can assist with beginning or continuing the process of contacting your inner experience and recognizing the role it plays in your life and therapeutic work.

The Conceptualized Self

The *conceptualized self* is probably the easiest to contact. It is the sense of the self that includes our histories (memories), beliefs, themes, explanations, reasons, categories, interpretations, and rationalizations. It is the coherent story we tell about ourselves when asked, "Why are you the way you are?" If we harken back to an earlier part of this chapter, you will recall that you can't know all the variables that shape the content of your life or that participate in influencing your behavior. You, of course, can remember certain aspects of your history and you have stories or verbal themes related to why you behave in particular ways, and even though these stories about yourself may be quite true, they are only a tiny part of the picture. Given that you don't have contact with all the variables that shape "you," hold any conceptualized self you are aware of lightly; it is not *you* anyway. You as therapist, you as novice, you as compassionate and understanding; you as incapable or capable, you as right or wrong—whatever the content, two things will be necessary for the self-awareness process. One is to recognize these conceptualizations as content—these are the experiences of your life. And second is to notice the attachments you might have to any of these experiences and how these attachments unfold in your current behavior (life and therapy). Do these attachments drive or inform you? And, are you willing to let go of these attachments if doing so works to help you move closer to values-based living? Are you willing to let go if any of these attachments is interfering with the therapeutic process?

Let's explore this a bit further concerning therapy. Standing back from the conceptualized self may free you from unnecessary defense or getting stuck in an outcome-oriented focus instead of process. It may free you from a too heavy reliance on metaphors and exercises or other rigidities that can arise in therapy. Let's look at what I will call an unnecessary defense. When we begin to make coherent stories about who we are, and hold too tightly to these beliefs about why we do what we do, then any questioning of these concepts becomes threatening. We begin to defend. An easy place to detect this defense is when a therapist takes a one-up position to the client. Here, the therapist, caught in their stories, rises above the client in some way. It might be, for instance, that the therapist identifies with their degree, holding their status as graduate or doctor in a prideful or even superior fashion. Getting caught up in these kinds of stories can be subtle or more obvious. Its subtle form might be in the power differential that can be present when sitting in a specific chair in the therapy room (e.g., the big leather chair facing the couch made of cloth, for instance). Noticing the function of sitting in that chair might be useful. Does it have a subtle impact on the therapy? It might. Being aware of how you show up in this chair will matter if it is about recognition of you as "*the* therapist." It will move you out of the position that ACT holds—we are not fundamentally different from our clients, we suffer too, and we, as much as they, should hold our conceptualizations of ourselves lightly.

Clinging to an attachment of self-as-therapist linked to an advanced degree can also lead to more obvious behaviors in the therapy session. Advising, explaining, and directing are all potential examples. Being aware of the meaning and felt experience of these kinds of exchanges is part of knowing oneself and the impact of your behavior on others. Inside of this power differential, compassion, empathy, choice, and a thoughtful implementation of the six core processes are each at risk. Attachment to content is routinely present for all of us and, undetected, potentially problematic.

REFLECTIVE Practice 3.2

Explore whether you have any stories that you are attached to that might interfere with your therapy. Consider both subtle examples (e.g., I deserve recognition for my hard work, I am a fraud or a fake, I am never appreciated) and more obvious ones (e.g., sitting in the "big" chair, having a story similar to your client's such as childhood abuse or divorce, routinely explaining to your clients that you know more than they do). To help get started, you might consider stories that interfere with or cause struggle inside of your personal life. These are more likely to be the ones that show up in therapy as well. We all have these attachments. Ask yourself:

- What thoughts and emotions arise as I explore these stories?
- Do I find myself wanting to defend, even in small ways?
- What would it mean to let go of these stories, to settle into uncertainty or to truly stand in a place where I am far more than any story I have about myself?
- What fears might arise?
- How might it change my therapy?

Take time to write about what you are exploring.

There is a full range of ways in which we can collapse into our stories. We can get caught by ideas, emotions, and sensations. Becoming aware of the stories that capture us and the way in which they potentially play a role in therapy is part of the work in

ACT. Indeed, we can even get caught with respect to becoming an ACT therapist. For instance, I am met with curiosity by individuals interested in learning ACT when I say, "ACT isn't the answer. Don't cling too tightly to it either." There are multiple reasons to make this statement, some of which we will turn to later in the book. But for now, being able to hold lightly "self as ACT therapist" may also provide the wiggle room to discover, to learn.

I once encountered a supervisee, new to ACT, who during our first group therapy session together interrupted me multiple times, leading the group into places that were not necessarily ACT consistent. When I asked about these interruptions in the supervision that followed, the supervisee simply stated, "I thought I had a better way of doing it" (I will return to the importance of my personal self-awareness in this moment later in the chapter). The supervisee, trained in psychodynamic therapy, saw opportunities for interpersonal process and took them without regard for impact or the previously stated plan. Perhaps in and of itself, the psychodynamic focus wasn't problematic, but he was there to learn ACT. He had collapsed into his thoughts about the group and a story tied to and organized by his previous training. He collapsed into his beliefs about what he saw happening in the group, "forgetting" that he was there to discover, instead clinging to his "better way" of doing it inside of his chosen theoretical orientation. In attaching too tightly to his perspective, he lost the opportunity to see what can unfold inside of ACT—a therapy he wanted to learn. Holding lightly may have proved useful under the circumstance.

Although it is useful to spend time focusing on the conceptualized self-as-therapist, it is also essential to be aware of how you imagine yourself no matter the context—as much as possible anyway. Sometimes it is difficult or challenging to know whether we are trapped in our conceptualizations. Working to understand how you might unnecessarily defend a sense of yourself is part of the self-knowledge process. For instance, you may come to defend your history, holding it as your literal self (e.g., I am my memories and experiences). If this is the case, and you have suffered childhood trauma for instance, then when you work with others who have childhood trauma, you may lose sight of the fact that they are not their trauma. You may begin to make interventions that support that story, perhaps inadvertently noting to the client that her experience damages her or that she *is* a victim and always will be. Whatever story you may be attached to about yourself, it could heedlessly appear in therapy, expressly if the client elicits in you the emotion, sensation, and thought experience associated with the story. Not that the latter won't happen, it will—we are not doing therapy in a vacuum. You respond to what the client brings, eliciting your history. When it does, however, what relationship will you have with that experience? One where you are pulled or compelled to respond, perhaps in a way that doesn't function to move the client forward, or one where you can notice and choose, responding in a way that is functional to the situation?

> ## Manuela's Reflection
>
> *To learn to notice when we as therapists establish a frame of coordination between us and our stories and to notice how that impacts our behavior in the room is very important. It helps us to be more flexible as a therapist. To be able to choose the stance and the perspective from which to work enables us to move to a broader contextual perspective of ourselves and thus our clients, which can be challenging. For instance, one of my more difficult experiences is to be able to "stay with" clients who show anger toward me. This sort of situation tends to make me fuse with a longstanding history of myself: I'm unlovable. When this happens, it is much more likely that my behavior will be about trying to fix things to make myself lovable. Engaging in these behaviors in the past was not always the wise choice, nor the choice that helped my clients. It was painful for me to notice that I was operating from that story of myself and then to take responsibility for this behavior. But it was worth it.*

It is in this place, from the ACT perspective, that knowledge of the self, awareness of content, facilitates process and relationship. The therapist's acceptance of their history, emotions named, sensations described, and thoughts had, frees them to model and convey this open and willing space. The therapist is not their story, nor is the client. Not bound by any story, choice is available. You don't need to rigidly avoid your personal experience in therapy, following a rule that therapy isn't about you. Nor do you need to share too much in terms of what is going on for you internally. Choosing according to function is possible. The very wish we have for our clients is there for us too. Someone once said to me, "Kill yourself every day." A lay reaction to this might include alarm; however, letting go, each day, of every content-oriented sense of yourself may not be a bad idea, notably if it serves you as you move in values-based ways, or if it serves your therapeutic work.

> ## Carlton's Reflection
>
> *I think there are a few problems with the instruction to "kill yourself every day." While it's undoubtedly an attention-grabbing phrase, specified in the form of a decontextualized rule, it doesn't take into account whether behaving under the influence of self-content is useful or not, and it seems to me that behavior driven by content is often beneficial. I wonder if instructions like this can, therefore, be unhelpful and confusing, and additionally, if it is possible to even achieve! Self-content, constantly created on a moment-by-moment basis, can never be stopped or "killed,"*

> *only paused on a very temporary basis. Surely what we are looking for is the ability to choose when to experience ourselves as content, and when not.*
>
> Robyn: The therapist and client are free to choose whether content will guide them. Choice is part of the hope or possibility in ACT. You and the client are free to choose. Unfortunately, when one is attached to content, the freedom is lost. Even content that proves useful. Context matters. And attachment to any story may cause problems depending on the context. Killing yourself every day is about letting go. It is about remaining flexible so that even quite valuable content in terms of values-based action can be set free when it is unworkable. When content becomes the driver of behavior, choice is narrowed. Clients will say things like "I have to do it this way" or "This will never end." Without observation of the content, seeing it for what it is, choice is lost. Therapists can do the same: "I could never do that in therapy" or "My way is the best way." Each of these, unexamined, can play out in unintended, values-inconsistent ways. Choosing is vital; awareness is part of choice, and holding content lightly, "killing it every day," will assist in this process.

Self-as-Process

The second sense of self is *self-as-process* (Hayes, 1994). Knowing as related to this sense of self is about being in verbal contact with ongoing experience. The fluid movement of emotions, sensations, and thinking, and awareness to these in the moment, involves seeing what is there to be seen as it is seen (Hayes et al., 2012). When we learn to bring awareness to our ongoing behavior, when we stay present to ourselves and our emotional experience, we can use it in relationships in healthy and positive ways. For instance, the therapist, aware of her ongoing experience, might reflect to the client her experience across a specific period during a session. The therapist might say to a client who never takes assertive action and once again didn't defend herself, "I noticed that when you started to talk about what happened to you, I first felt anxious, but then I began to notice a sense of rising frustration. I noticed that I had an urge to take action, to stand up for you, where you didn't stand up for yourself. Do you ever notice frustration or the urge to stand up for yourself?" Here, this awareness of self-as-process can be used in collaboration with the client to be "co-aware" of self-as-process. If the experience is similar for the client, frustration and the urge to stand up for herself may be signals to make discriminating choices (e.g., be assertive when it is workable). This kind of response is quite different from simple reflection or canned empathy (e.g., "That must have been hard for you").

Engaging in knowledge of self-as-process is largely trained, a part of our learning history. However, there are those who may have learning histories that might counter this kind of knowledge. For example, if you were often invalidated as a child, growing up inside of the historical "community" of an invalidating parent, you may not know the experiences inside of self-as-process. Your experiences may not coincide with those of the present-day verbal community. This difficulty can potentially show up in a few ways, including not knowing what you feel, demonstrating body language that corresponds with one emotion but reporting another (e.g., angry body language while communicating calm), and communicating the same experience across multiple body language and sensation experiences.

If part of a healthy and responsive psychotherapy is about being able to report experience in the here and now, the relevance of self-as-process is immediately evident. If, as a therapist, you are unaware of the ongoing experience of self, modeling this for clients who struggle with knowing their own experience will be challenging, if not impossible. This lack of awareness is where, I would stipulate, that mindful practice, as well as other awareness strategies, becomes an essential part of the ACT therapist's practice.

People who are new to ACT, and even some seasoned therapists, question whether mindfulness is necessary to the delivery of ACT. The answer is no, you can choose not to practice. But I would like to make an argument for yes. If we merely look at self-as-process as a vital part of being open to what rises and falls, to witnessing the coming and going of experience and from that place recognizing the freedom from the threat of internal experience, then awareness to ongoing experience in the moment should be fostered as part of ACT's foundation. As ACT therapists, we generally speak about this work, being aware of the flow of experience, in relation to assisting clients. I hold that it also aids the therapist, not only by increasing their capacity to pay attention in the moment and to mindfully listen to clients and understand what they may experience when asked to do mindfulness themselves, but also by cultivating personal contact with self-as-process. Awareness to self-as-process will help clinicians to identify and weaken the social contingencies that lead to a preservation of a conceptualized version of the self (see Hayes et al., 2012), growing personal flexibility.

When engaged in the ongoing awareness of experience in the moment, the historical and future conceptions and the judgments regarding those "loosen." This awareness of experience is potentially relevant to the therapist in session in many ways. First, it invites freedom. The therapist can be flexible in the moment, responding to what is currently happening rather than a concept of what should be happening. It allows the therapist to meet the client in the here and now. Suppose someone has asked you to work with a client with borderline personality disorder. Notice, right now, your conceptualizations of this person and how you might already be judging the client before they have stepped into the room. Self-as-process loosens you from these conceptions and judgments, giving you more freedom to respond to the client naturally, rather than reacting to the categorizations and conceptualizations of the person.

> ### Carlton's Reflection
>
> *The key for me is awareness; if I am aware, I have choice. I want to have the freedom to have both a conceptualized self and a conceptualized other, and even to make and fuse with judgments, if doing so is useful in a specific context.*
>
> Robyn: *It is the case that awareness and choice matter. But also, engaging an ongoing awareness of experience and choice is part of the process. Looking to see how a conceptualization is useful may be important. However, fusing with those conceptualizations is different from being aware of them and then choosing to act on them. If fused, how do you know if it stops being useful, or if you are able to let go? As well, if fusion impacts your life and your therapeutic work in ways that stop you from engaging your values or that lead you to stigmatize your clients, treating them in ways that are problematic or impulsive, such as reacting to a judgment (e.g., this client is too difficult), are you able to flexibly respond and adapt? Choose to hold if you choose to hold and it makes functional sense, but hold lightly, so you can choose again if needed.*

Second, this awareness of process invites flexibility in another way. If you are aware of process, you are less likely to hold on or impulsively respond to any judgments you have about the client overall, or in a moment in session. If I persist in practicing self-as-process awareness, then I can acknowledge any experience that rises and falls in the moment and not necessarily respond to it. Here's a personal example. I worked with a client who, time and again, noted his superiority to those around him, through vague reference and hidden statements. I conceptualized his interpersonal relationship problems as tied to this inflated sense of self. At times, I found myself judging him and making mental notes about how difficult he was to like. He once came to the session and reported a profoundly shaming experience wherein someone had confronted this inflated sense of self in a very public and humiliating way. I collapsed into my judgment of this client—buying my concept of his self-centeredness—and inadvertently and impulsively sided with the person who had done the humiliating. The client balked at my response and left the session feeling even more alone. As I reflect on the experience, I see that my conceptualization of him took precedence over my own in-the-moment experience, which was a sense of confusion and feeling quite sad. Had I slowed down and attended to my internal experience (my thoughts and attachment to them, my emotions), I might have provided a more attuned response. I might have empathized with the pain of his humiliation and either waited another day to work on how he relates to others or found a more tender way to see what might evoke this confrontational behavior that he experienced from another.

Third—and this is a more personal hope for all therapists, and indeed humanity—the richness in life is not found simply in what has happened or what might be, but in the ongoing flow of emotion, sensation, and thinking. We are beings in motion. The fertile and abundant quality of experience is what makes life so romantic, powerful, and full (including what unfolds in therapy). Awareness to that process in a routine way may be the very definition of life and part of what makes it vital.

Beyond my hope for therapists becoming aware of their ongoing process in life and therapy, there is also some interesting data that suggest that being mindfully aware may improve or influence the therapeutic relationship and client outcomes. For instance, Fatter and Hayes (2013) found that meditation experience predicted the therapist's ability to manage emotion in session effectively. Ryan, Safran, Doran, and Muran (2012) studied the relationships between therapist, dispositional mindfulness, therapeutic alliance, and treatment outcome. They learned that therapist mindfulness, based on the extent to which they act with awareness (as measured by the subscale of the Kentucky Inventory of Mindfulness Skills; Baer, Smith, & Allen, 2004), positively correlated with clients' ratings of the client-therapist working alliance. Ryan and colleagues argued that dispositional mindfulness may be an important variable in psychotherapy outcome. As well, the Act with Awareness subscale of the Five Facet Mindfulness Questionnaire (Baer et al., 2008) indicated a positive relationship between patient-rated working alliance and improvement in interpersonal functioning.

In considering mindfulness practice as it relates to cultivating awareness of self-as-process and its role in creating openness, I want to consider the different possibilities and aspects of the practice. First, *dispositional*, or *trait*, mindfulness refers to the level of mindfulness a person has during everyday activities, as opposed to *state* mindfulness, which is the level of mindfulness a person achieves during, or after, engaging in mindfulness meditation exercises (Cahn & Polich, 2006). Attunement to what is going on inside the parameters of your skin—being aware of mind and body—can be developed and engaged in numerous ways. A regular practice linked to both trait and state is essential. This practice can range from participating in more formal daily practice to attending a mindfulness-based stress reduction course to movement meditation like tai chi and qigong. It can include using downloaded phone apps that guide you through mindfulness to simply paying attention to small things like water falling on your skin while showering or the dirt in your hands while gardening. How you as a therapist cultivate mindful awareness to your ongoing sense of self-as-process is nearly endless. I speak to this range of activities to harken back to the question, "Do I have to meditate or practice mindfulness to get this [ACT]?" If you want to more fully contact the ongoing flow of experience in the moment and use this kind of awareness in your therapeutic work (being able to notice the challenges that mindfulness can pose, but also being able to experientially contact an open, defused, and accepting present-moment experiencing), then I would argue, as noted, the answer is yes. But do it in a way that works for you. Try a bunch of different things. Build your attunement through a flexible mindfulness practice. Take a little time to listen to yourself, tune in. Carefully observing your experience during formal meditation or other mindfulness work will assist you in actively choosing your responses to the world (and in therapy), rather than getting whipped around like a chaotic wind by problems of the day or habitual reactions to stimuli.

REFLECTIVE Practice 3.3

Take a few moments to reflect on your mindfulness practice. Explore these questions:

- Am I aware of my own level of mindfulness during everyday activities (i.e., dispositional, or trait, mindfulness)? Is this an area of personal growth for me? About how much time do I spend in awareness to activities of the day?

- Do I have a formal mindfulness (i.e., state mindfulness) practice? If not, what is the barrier? If yes, how much time in a week do I devote to this practice? Am I in a good place or would I like to grow this practice? What am I willing to commit to?

- When in therapy, what does my mindfulness look like? Am I able to be present to the client and be aware of my internal process? How might working on this benefit me?

Self-as-Context

The third sense of self is *self-as-context*, or *perspective taking sense of self* (see Hayes et al., 2012). It involves contact with a sense of wholeness, transcendence, or presence. This sense of self is the experiencer of internal events, not the events themselves. Pure conscious awareness might be another way to describe this sense of self. It is here that we have full liberation from specific content and all broader concepts, including concepts such as "I" or "self"—these too are a set sounds that refer to something; they are not the something itself. It is from this perspective that no content, no matter how painful or important, defines the being.

Carlton's Reflection

I have heard a number of people, both clients and therapists alike, comment that experiencing a sense of the self as a context is a very elusive experience, and indeed this matches my own experience of meditative practice. It seems to me that the

> *liberation from the concept of "I" that you describe is not going to be possible for the majority of clients to achieve over the course of ten to twenty therapy sessions.*
>
> Robyn: *Perhaps not possible and maybe even not necessary. However, seeing "I" as a word, as a sound that refers to something, itself may not be too challenging for either a therapist or client. It is, however, particularly sticky. It is hard for us to know that there are billions of "I's" on the planet, as we typically see "I" from one perspective. I see, I feel, I think. Contacting the process of observing seeing, feeling, and thinking from a position of awareness is part of the work done in therapy, whether done within a few sessions or more. It is challenging though because observing the "I" can loop back on itself—it's like having an awareness of being aware, i.e., Who is observing "I"? I am. Observing the being, or the experiencer, who is saying "I am" is also part of observing ongoing process. Defusing from "I" might be useful, but not necessary. It simply depends on workability. Nevertheless, contacting self-as-context includes the recognition that even "I" is content. From a more personal place, recognizing this has helped me to connect more fully with being itself. I am not an "I"—I am consciousness. I am being. And if I could find a way to distill this experience into words, I would. Even saying "I" in the previous sentences reveals the difficulty. Awareness to simply being, full stop. As well, when I recognize I am not separated from others because I speak from the position of "I," I can connect to a larger whole. We are all beings experiencing, interconnected through interaction with context, whether it is earth, animal, or human.*

Here too, lies an incredible amount of freedom. Any sense of you as conceptualized is neither threatening nor favorable. You are not defined by that which is categorized by bad or good. Here you are free from attachment, and thus suffering.

Contacting this sense of self permits the freedom, or flexibility, to view or take perspective on the many roles and self-conceptualizations that we hold. Operating from a broader perspective, transcending all notions of what defines us, offers the opportunity to explore many different aspects of our experience and, to the degree that we can, aspects of others' experience. It is this capacity to move in and out of our conceptualized senses of ourselves, while remaining intact beings, that is useful for empathizing, connecting, and flexibly responding to the environment, context, and client.

Let's return to the earlier story wherein a supervisee had told me, "I thought I had a better way of doing it" when I asked why he had made so many interruptions during our ACT group session. As noted, I had suggested that he had fused with the thoughts he had about what was happening in the group. Through a series of questions, I learned that he thought that he "knew better" than I about what to do and that he had read about ACT and therefore understood it "quite well." You can imagine this supervision experience turning out a number of different ways, and it would be inaccurate to tell you that his behavior wasn't pushing my buttons—it was. I was observing all kinds of judgment and feelings of frustration. It was in this place that my awareness of my own in-the-moment experience, as well as my nonattachment to my own conceptualized self as a "seasoned" ACT therapist was especially important. This is not to say that I didn't have brief moments of fusion with my judgments; I was moving between being caught and not. However, through awareness and nonattachment, what unfolded led to one of the more enriching ACT supervision experiences that I can recall. I made a choice: rather than respond impulsively, I paused. I then gently asked, "Do you have a sense of how this experience might be impacting me?" He said, "You might think me arrogant." I responded honestly but not from a one-up position, "Yes, that thought crossed my mind." It was somewhere right in there that I did some perspective taking. I put myself in his shoes. What I didn't sense was arrogance, but rather a desperate clinging to a sense of self designed to avoid fear. So, I pursued a brief line of questioning about that (e.g., "Have you been accused of arrogance in the past?" [yes]; "What would happen if you couldn't rely on 'knowing better'?" [insecurity, fear]). It was in this open, unattached place that I was able to shift and move into something more useful in our context. Instead of a battle of wills (a battle of one conceptualized self fighting with another), we stepped into the very murky water of fear and began to swim.

I give this example to illustrate both the potential problem of attachment to a conceptualized self as well as the freedom created when you hold your conceptualizations lightly. There are examples without number (and I am no saint; I get attached). The key is being able to shift flexibly. The flexible shift from myself as more than ACT expert to my emotion and thought experience in the moment, as well as taking the perspective of the supervisee, allowed the two of us to work inside of a more authentic process. We explored his fears and avoidance, my sense of needing to defend, my capacity to see him as more than arrogant, and his capacity to see himself as more than insecure. An effective and ongoing supervision emerged. To be sure, I am not always on my toes about these kinds of things. If you recall, I collapsed into being right about my conceptualization of a client's behavior and created a truly failed session when I missed the opportunity to empathize with the client rather than with the one who humiliated him. This work is a process, not an outcome.

There is a full range of ways in which we collapse into our interwoven ideas, emotions, memories, and sensations that form our conceptualized selves—and the "self-as-therapist" is no exception. When overly attached to self-as-therapist, for instance, we can slip into "knower" and "not knower"—we can inadvertently lose our humility and move into a one-up position with the client or perhaps cling to the notion of what it means to be a "good" therapist, thus not being willing to be vulnerable, make mistakes, or try new things. The intricacies and interrelationships among private events; how you talk to yourself about you, others, and the world; as well as the ideas you have about others' ideas about you are all there to be dispassionately observed. You are more than these things.

REFLECTIVE Practice 3.4

Take some time to consider how you might gain greater self-knowledge and become more present to self-as-context. Several approaches may assist you, including increasing your mindful awareness practice and using supervision and consultation to discover or explore your attachments and defenses concerning therapy.

If possible, bring willingness to this circumstance in discovering how your view of the world, self, and clients influences your work. Notice any ways in which you might be buying a sense of yourself, and ask yourself these questions:

- Is this sense of myself useful in the context of therapy, or in the context of my broader life?
- If I am more than this sense of myself, what does that free me to do?

Making ACT Personal: Owning It

From the place of nonattachment, we can return to the chapter opening: Given that I am part of the context, *my* behaviors and their influence on the client are also worthy of consideration. How does my behavior and what stance I take in therapy influence the experience of the client, the session, the context? If it is the case that we contribute to our suffering through attachment or clinging to notions of ourselves, might it also be true that clinging and attachment play out in many ways, including inside the therapeutic context? We can lose the liberation that self-awareness and perspective taking can create when we are unwilling to explore our attachments and conceptualizations; falling into an overly mind-y place will influence how we perceive ourselves, the rest of reality, and therapy. The lens we look through may set us up for losing contact with the client as more than their concepts, emotions, memories, and sensations. We may inadvertently make mistakes here, not only by buying a concept about the client but also by buying a concept about ourselves. When we collapse into concepts, we lose contact with the broader sense of self and risk our ability to accept ourselves and our clients as is. Learning to come into the present, connected to a larger sense of self, so that we are no longer slaves to age-old concepts, is the antidote to the attached condition. Life only happens in the here and now. I/here/now is all there is.

> ## Carlton's Reflections
>
> *Is it not the case that life also "happens" in our pasts and in our futures? It seems to me that adopting these perspectives can also be useful, as demonstrated by the I/there/then of the personal examples you have described in this chapter. In this sense, I/here/now is maybe not all there is.*
>
> *Robyn: Agreed to an extent, I/there/then has been useful, but I can only tell it from I/here/now. And whatever I might say about what is coming is a conceptualization of that future. And I am speaking to it from I/here/now. I hear what you are saying, and these conceptualizations of past and future are important. I would hope that nothing I have written would lead anyone to believe that these aren't a part of our experience, a part of how we relate to ourselves and others. Indeed, even values are conceptualized, and they guide me in the meaning I want to create for myself as I move forward. But only "now" is what I have.*

The Heart of Now and the Door to Next

Given that therapist reactions to clients are inevitable, and that client and therapist are each continually affecting and influencing the other's internal and external experience, it is the therapist's task to use this information wisely, overcoming vulnerabilities to respond impulsively, instead working from a more flexible place—an open and aware place. From the position of self-awareness, we recognize our efforts to control (e.g., to get it right, to perform, to do it like others, to understand), our places where we are attached and collapse into different levels of rigidity. It is through routine practice in being open and aware, in connecting to a larger sense of self that letting go of attachment is possible. Flexibility lies here, both for your personal choices and for assisting your clients in making theirs. You can freely define and choose concerning values in creating a meaningful life. It is this door that we walk through next. In chapter 4 we explore existence and purpose. We look to the role of death in creating meaning and engagement in our lives. Death is a catalyst. Will we face it and allow our awareness of it to influence personal change?

CHAPTER 4

Engaged: Existence and Purpose

Death is no pretense. It is as stark a reality, as complete a presence as is life itself, the other ultimate choice.

—Karen, client of Yalom

The exploration of existence and purpose, from my perspective, is fundamental to ACT, the therapeutic stance, and engagement in process. The reality of our own death and the responsibility to make ongoing, purposeful choices are central to our work in acceptance and commitment therapy. Practicing choice from an open, aware, and nonattached presence allows us, in the inevitable movement toward death, to create our meaning with intention. Indeed, awareness of death has the potential to instigate a radical shift in life perspective and motivate us to engage in being alive in the moment and commit to actions that serve our values. Awareness of death can move us from a state of wondering "why" we live to a process of engaging "how" we live. It catalyzes change. We recognize that death isn't seeking to be our enemy, it is not an entity with sickle and hooded cape coming for us; it is simply a part of life. Undoubtedly, it can be an ally in the pursuit of meaning. As such, part of our work in therapy (and in life itself), from a kind of sacred or acknowledging stance, is to facilitate awareness of death as an ally in developing a sense of choice and responsibility for personal meaning and purpose. It is here in the awareness of death that a genuinely authentic immersion in life can emerge.

In this chapter, I will briefly explore existentialism and its interconnectedness with ACT. I am unable to cover all the interests and intricacies of ACT and existence, as that would take a book of its own. The chapter will not as much focus on client issues as on the broader subjects of our own existence and its meaning. Nevertheless, I invite you to reflect on your personal experience as you read: to survey your own relationship with death and its link to meaning; to consider how you bring knowledge of death into your own life and into your therapy. As you read, notice your reactions and wonderings. I invite you to be curious about how existence and its inexorable conclusion impact your interpersonal and intrapersonal experience, processes we will turn to in part 2 of this book. What role does this knowledge play in your life and your therapy? Explore the passage of time and its ramifications, and how it might lead one to consider the importance of process over outcome, both in life and in therapeutic work. As for each of us, there is only one true arrival: it is death.

> ## Carlton's Reflection
>
> *Of course, for many people, death is not actually viewed as a finality, as an end to existence, as the ultimate arrival, but instead is understood as a point of transition to an afterlife. From your experience of working with clients who hold religious and spiritual beliefs, does this change how you might use the possibility of death to bring the meaning of life into sharper focus?*
>
> *Robyn: Many, if not most, do not see death as final. Spiritual and religious issues are important and can support the work we do concerning values. As therapists, we have for too long ignored the spiritual, relegating it to clergy. I believe this a mistake. Spirituality and religion are essential to clients and part of their suffering and joy. Speaking to transition may be helpful. Yet even with belief in an afterlife, fear of the transition remains. Doubt and uncertainty about what follows death still create existential angst. The therapeutic work remains to be done.*

Death

My mother died of cancer on a September Tuesday in 2008. As she took her last breaths and settled into a final exhale, my inner world turned inside out. The pain of the moment was too big to describe, too immense for words. I was mottled with grief, paralyzed in the moment. My existence questioned, I felt completely undone. And then another thing happened. The world kept turning. The clock still ticked. I was in more disbelief that time had not stopped than I was of her passing. She was gone; the world should end.

In the days after my mother's death, I continued to marvel at the strike of the clock and the rise of the sun. It was a stark reminder of motion, the impossible stoppage or irreversibility of time. It was also a stark reminder of my own finitude. Her death brought me into contact with my own—I will die. And time will move on. The death of those we love or of the creatures we care for, aging, or simple milestones such as birthdays and anniversaries can place each of us in contact with our own passing. This is at once a terrible and beautiful thing. Recognition of death can bring anxiety and fear, but it can also bring creation and purpose. Death's qualities—the unknown, "nothingness," the cessation of our consciousness, of our existence—can be frightening. However, recognition of death can also be freeing. The Stoics of ancient Greece pronounced, "Contemplate death if you would learn how to live."

Reflecting on our finitude is not a morbid endeavor as some may fear, although it may contain anxiety. Anxiety is a part of existence, and existence cannot be postponed (but by one single act). There is no waiting to exist without anxiety. Waiting to feel different is existence, too. A life dedicated to sheltering from difficulties, freedom from fear, and relief from pain is still a life being lived. Even in death itself, there is existence. We live until the last moment, experiencing. Reflecting on our own death then is in the service of living consciously. It is in the service of fulfillment and meaning. It is an encouragement to be aware, to every extent possible, of each moment of existence. As George Santayana (1923) put it, "The dark background which death supplies brings out the tender colors of life in all their purity."

Death and Clarifying Values

Defining and clarifying values in ACT is in the service of creating meaning; it is part of these tender colors of life. But these colors are inexorably linked to death. We have a limited time to engage our purpose, to touch the colors in the moment and create them in our steps. I believe that explorations of meaning will necessarily be explorations of death. Therefore, your and your clients' exploration of values will be about the bittersweet dance between engaging a full and vital purpose and the passage of time, aging, and death. This is not always an easy task. Exploring our own mortality, let alone our clients', may feel or seem difficult. For example, I have seen avoidance of death in ACT work itself. The Funeral Exercise (see Hayes et al., 2012) invites us to visit our own demise. We are asked to attend our own funeral and consider a eulogy. Therapists being trained in ACT have asked in response, "Can I skip the Funeral Exercise or do something else instead? It seems too morbid to me. I am worried it will scare or offend my client." And, indeed, there are many exercises one can use to tap into values, other than this specific exercise, and through doing these alternatives, one can avoid explorations of death. In so doing, however, I worry that we are conspiring with our clients and perhaps avoiding our own fears and anxieties surrounding this issue.

Nonetheless, there is no exercise or reflection as potent, perhaps, as contacting one's ending. In confronting our own passing, we are better able, and more willing, to help others see the power of awareness of that which is inevitable—death—helping us to wake up to the tender colors of life and personal creation. So we must see it ourselves and not shrink away. For if death is too hard, too scary, too morbid, then what motivates us to engage our values today? Why change anything now? We do not know how much time we have. In contemplating our own end, we recognize that there is still time for life in whatever time is left. Values are there to be lived until that last inevitable moment. To acknowledge this is freeing for you and your client. The work here is dynamic, curious, uncovering, not morbid. Recognizing possibility lives right inside the awareness of our end. We have this one lifetime. The capacity to turn and take a different choice is available as long as we are aware and alive.

Ordinarily, however, we slip out of awareness, our thoughts taking us to many places that are not about the immediate beauty of the world or the experiences of seeing, touching, smelling, or tasting. They take us away from kindness and love. Our thoughts keep us in what we cannot do. Awareness of experience is dwarfed by small concerns and ongoing fears, by entanglement in worry and the past. Presence in the world is taken over by threats to our pride, our egos, or our concepts of ourselves. It is in this recognition of the easy slip from awareness that the valuableness of practice and process emerge. Returning to awareness, over and again, in the ongoing process called "life" is essential—your personal practice of awareness again revealed.

Carlton's Reflection

This is certainly my experience. When I contemplate my own death, it focuses my attention on living; it pulls me out of a preoccupation with the minutiae of the day-to-day and gives me a perspective on life that can sometimes feel imbued with meaning. However, and very much contrary to what I would wish, I am aware that such moments are often too fleeting. I find it very difficult to hang onto that perspective for a sustained period of time; before long I am absorbed in the minutiae once again. The change in perspective is never permanent, only ever a movement between one viewpoint and another.

Manuela's Reflection

Talking about death in therapy may sometimes be taboo, something we all implicitly or explicitly agree not to do. But I learned in my years as a therapist that when we personally dare to talk about mortality and our own death, we are able to encourage our clients to talk about it too. Meaningful work emerges. Fundamentally, there is a

> *quality of being human and vulnerable that is carved under the presence of death, tending to make us humble and grateful. The uncertainty about not knowing when our own death will arrive may also teach us to shift from futures that we may never know to values and process, moving us to the moment-to-moment experience of meaning and purpose. Notice, from an authentic place, the felt sense of your own mortality. Be aware of what this recognition brings to you regarding meaning and purpose; then open the door for your clients to do the same. It is truly an adventure.*

REFLECTIVE Practice 4.1

I invite you to explore your own death. Take a period of time, a more extended period, a half hour to an hour, to quietly sit and reflect on your own passing. Let yourself fully show up to your personal end. Perhaps imagine darkness, or never waking from sleep. Imagine those whom you love without you or no longer being able to engage the world. Dig deep; see if you can stay with a full acknowledgment of your death. Notice what happens as you explore this place. Be both curious and examining. Be aware of your fears and anxieties as well as your joys, if they show up.

After you sit, explore these questions:

- What did I notice? Or experience?
- As I touched this space, did I find that there is anything I am pretending to not know on a regular basis?
- Do I think I have time, or does my conception of a long-lived future deceive me?
- If I had only a year, what would I be doing? Am I doing it now?
- What scares me about death, and how might it interfere with my life? With my work with clients?

This exercise may be part of the beginning of personal discovery or a continuation of something you have already explored. Whichever the case, don't stop here. Keep in touch with this process, not as a morbid exploration, but as a dark background wherein death can bring out the tender colors of life and meaning in all their purity.

Death and Self-as-Context

Notice the process of describing yourself. I am a therapist. I am a client. I am thoughtful and engaged. I am lonely and unlovable. I am a hard worker. I am lazy and bored. I am quiet and small. I am loud and angry. I am a man. I am a woman. I am young. I am old. I am smart. I am okay. I am cool. I am stupid. I am unworthy. I am worthy. And on it goes, your mind defining you. Your core lost. Your essence, your "beingness," stolen. Part of the self-as-context work in ACT is about reaching beyond mind to pure consciousness by shedding over and again thought, desires, hopes, and dreams. But it is the awareness of death that can assist this shift in perspective. We encounter change across time in our own aging, a fierce reminder of death. As our consciousness stretches across time, we see change, we see the ongoing flow of life. It is possible for an individual to let go when emotion and sensation are encountered and experienced by a being, but are not the being. Fear is a wave, flowing over one's surface; it is not the surface itself. Sensation rises and falls; it does not disrupt the being experiencing the rise and fall.

Holding all concepts of the self lightly, stepping in and out of each in the service of meaning, is the work of a values-based life. By recognizing a sense of you that is more than any concept, by letting go of any construction of the self and the world, choice is possible, freedom is alive.

REFLECTIVE Practice 4.2

Letting go of conceptualizations can be enhanced by contacting death. Sit in a quiet space and do the following. Against the background of your own death, imagine the least favorable part of yourself. The one who is afraid and intimidated. The one who is shy or perhaps easily taken advantage of. The one who is petty and ego driven, wracked with pain and anxiety. Whatever the least favorable part of yourself is, let this sense of you be present. Reflect upon the qualities of this sense of you. Let yourself touch the emotions and thoughts wrapped up in this aspect of who you are, have been, and will be again. And as you notice this sense, imagine that something has happened, some strange twist of fate, and you can no longer contact this aspect of yourself. It is completely swept away. Notice if there is anything you cling to, anything you might miss. No matter, this isn't you anyway. You are more than this sense of yourself. And what would it take to let go? To hold this sense of you lightly?

Against the background of your own death, imagine your "best" self. The one who has courage and stands strong. The one who is smart and successful. The one who laughs and is full of fun. The one who feels ease and peace. Whatever your best self is, let this sense of you be present. Reflect upon the qualities of this sense of you. Let yourself touch the emotions and thoughts wrapped up in this aspect of who you are, have been, and will be again. And as you notice this sense, imagine that something has happened, some strange twist of fate, and you can no longer contact this aspect of yourself. It is completely swept away. Notice if there is anything you cling to, anything you might miss. No matter, this isn't you anyway. You are more than this sense of yourself. Would you be able to let go? This isn't you either.

Remember, any conceptualization held too tightly limits your freedom. Even the ones you quite like.

Freedom and Responsibility

I use responsibility here to reference our capacity to respond. We are the author of what we do now and next in life. We are responsible for what we create, our own destiny, our personal meaning. Fully connecting to this understanding can be enlivening, helping us to live more boldly. However, it can also be quite frightening, and for many, it feels so scary that they deny this level of responsibility. However, the denial doesn't change the thesis. It remains an existential concern. The question "What will I do with my existence?" still applies.

Heidegger referred to the individual as *dasein* (Wheeler, 2011). The individual is "there" (da), the individual exists as an object in the world, and the individual is "being" (sein). The latter is transcendental or aware; it is conscious of its "thereness." It is both aware and there. This inevitably leads to responsibility. Aware of your being, what will you create? Inside of being able to observe your "hereness," what will you do? In this fashion, responsibility is ultimately linked to freedom. You are here and free to choose. You are responsible for the life you create. As well, based on the perspective of existential philosophers such as Sartre and Heidegger, we are also accountable for the attributions we give to the world. Languaging behavior becomes incredibly important here. In recognizing the arbitrariness of words, we can also find freedom. Growing our awareness of words as something we humans have devised, not actual things that exist in the world, we arrive at both meaning and meaninglessness.

We have constructed the world. The flower does not know that it is yellow or red, and the bee does not know that it bumbles. The elephant does not know that it is an elephant, nor that it is majestic, or close to extinction. The bird does not know that it flies or that it is a predator; nor does the ant know that it is small and industrious. The sun does not know when it sets; nor does it know that it is beautiful. The moon does not know that it is blue, or that it has been walked upon by man; nor does it know its location concerning Earth or the universe. Indeed, the universe does not know that it exists. Time does not know how to keep itself. In this same fashion, the emotion does not recognize that it is depressed or anxious. The painful sensation does not acknowledge that it is challenging or bad. The heart does not realize that it is broken. Nothing has significance in the world but by our creation. This doesn't mean that language is arbitrarily applied. We use language inside of social learning systems that inform our application of the same. However, standing inside of the space of awareness that nothing has significance in the world but by our creation, what will you choose? Meaning or meaninglessness? It is here that the emphasis in ACT on choosing your values is essential. Values are largely constructed. With this awareness and the awareness of being, how will you respond? What will you do with this freedom? What will you invite your clients to do with theirs?

In Sartre's view, human beings are not simply free, but doomed to freedom. One is wholly and entirely responsible for one's own life, not just for the behaviors chosen and acted upon, but also for the failure to act. This, however, is not a moral issue. It is not a matter of what you *should* be doing, but rather a matter of being entirely responsible for what you *do* do. This knowledge is both liberating and frightening. To know that you can choose something other than what you have chosen up to now might bring with it possibility, as well as fear. You can observe this by imagining a simple action. If you were to lie this book down, stand up, and walk away from the life you have today, simply leaving it behind, choosing a different direction, you might be able to feel the possibility (e.g., living on a beach in the Bahamas, working less, engaging in social issues) and the dread (e.g., leaving behind what you have created so far; leaving loved ones, places, or jobs). What is vital in this current exploration is not merely what you choose, but *that* you choose. You have the freedom, you are able to respond, and you are responsible. What you choose is only relative to the meaning you would like to create. You are here, you have as much time as you have, so what will be your meaning?

Note that awareness of responsibility is not synonymous with change. You must commit yourself to some action for change to occur. As well, action extends beyond the self. It involves interaction with the environment and the interpersonal world. You must move. Your feet must be engaged in forward motion while you take responsibility for the direction. And as you step, the path of meaning will unfold. Therapeutic change is also expressed in action; awareness of or insight about freedom and responsibility alone will not create meaning.

If we are ultimately responsible for our lives, it then means that as therapists we need to turn responsibility over to the client. It is worth examining your own stance concerning freedom and responsibility so that you may arrive at a consistent position. The more active the therapist is in taking responsibility for the client, the less likely the client is to engage it. As a therapist, "have to's" and "musts" are relinquished to choice. The client need not do or change anything. It is up to them. I consistently turn responsibility over to the client. This can be challenging, as the relationship between the environment and personal choice is extraordinarily complex. Reciprocal processes are at play, and consequences are there for every behavior. It is why process and continuous choosing in the moment are essential to the creation of personal purpose.

Manuela's Reflection

Recognizing client responsibility was truly a turning point for me in ACT and in doing psychotherapy. I tend to do a lot of work to improve my client's life. I want to

> *help them make their life better. I used to take responsibility for the client's life and put in more effort than they did. This was a mistake. To recognize that clients are truly responsible for their lives, that they choose to create, was, for me, a milestone. It made me humbler about my work and shifted my perspective and therapeutic stance. When I find myself taking responsibility for my client's life, I remember Robyn's words in supervision. I repeat them to myself like a mantra: "Give your client's life back to the client." After this recognition and change, I was much more willing to linger in "choice spaces."*

REFLECTIVE Practice 4.3

I invite you to find a quiet place where you can meditate and reflect. Get a blank sheet of paper and a pen. Draw a straight line on the paper. Mark it so that one end of the line represents your birth and the other end your death. Draw a cross on the line to represent where you are now on this time line. Meditate on this for five to ten minutes.

Though you can be drawn into considering your own death, notice how quickly you move on after the exercise. Be aware of the rapid fade into busyness or denial, and see, in this fast fall back into remembering that you still exist, if you can choose where next you will place your feet. Then meditate upon your death again, and repeat the process of noticing described above.

Compassionate Immediacy

Personal specialness is a myth. Brief moments of consciousness of our own death, apart from the small recognitions that seem to force us into denial or fear, push us up against the realization that we are one among many (indeed billions). Just like those around us, we are finite and really do come to an end. My crushing realization that the world will continue to turn despite the death of my mother, that it will continue to turn despite the death of family and friends, and that it will continue to turn despite my own death, is a bit of unpalatable truth. I am not special. (But I am also not alone.) This truth carries an immediacy, or what I call "compassionate immediacy"—entailing two main elements: a sense of urgency and a sense of purpose.

Urgency

I cannot know when I will die, and my time here is short regardless of whether I could have this knowledge. People near death note that if they had only known, truly known how quickly life moves, if they could have simply connected to an awareness of their own smallness in the magnitude of the world and the universe, their own nonspecialness, then they would have lived life differently; they would have acted more boldly. *Now* is the time to choose.

In putting this down on paper for you to consider, I want to express a sense of urgency, for you and your clients. This isn't a frantic experience or an unpleasant pressure. Instead, I want to convey with heart and presence that time is running out, as John Donne so poetically expressed in "For Whom the Bell Tolls." I wholeheartedly hope for my clients then, to grab life here and now. I hope for them to live engaged, with vitality, experiencing the richness that life holds, its ups and downs, its cruelties and utter magic. I hope this for you as well. Choosing to step forward into creating meaning means that all that life has to offer will arrive—pain and joy. Again, you are invited to consider your stance concerning this recognition: your clients will die. Will you stand in compassionate urgency?

As a part of this urgency, I tend to "lean in" on creative hopelessness (a concept we will turn to later in the book), diligently working to undermine excessive internal control. Not only does control of internal experience lead us into paradox (e.g., the struggle with pain begets more pain), control also gives us the illusion that we can stave off death. If we can be immortalized with success, if we can be acknowledged, commemorated, or memorialized by what we do, then we can be preserved. Spared our own demise. If we can hold off fear and anxiety, move away from dread or tears, control our thoughts and sensations, then we can avoid the final truth. If we are continually projecting ourselves forward into the future, conceptualizing what will be (reworking or rewriting what was), then we can escape what is. But this very process of control—the very behaviors used to manage our experience—creates a space where actions are driven, fixed, and inflexible. Creation and possibility are lost, curiosity and openness abandoned.

Steven Hayes has on occasion asked ACT training participants to recall the first names of their great-grandparents. Turns out to be difficult—and for some, nigh impossible (if they even knew their names in the first place). We are not remembered for long, even by our relatives. We are not special. One of my favorite stories from Irvin Yalom (1980) is about a time when he was sitting on the beach in a lounge chair studiously reading. He looked over at a nearby bartender who was simply leaning on the counter looking out at the ocean, doing nothing but gazing at the sea. He recalled being a bit proud of himself and his work to get ahead as compared to the bartender who was "lazily" leaning on the counter. It was then that a question occurred to him: Work to get ahead of what? We have a short time to be here. Might we show up to now?

Awareness of death moves us into the importance of immediacy of making choices in the service of meaning-based action. We are simultaneously being and doing. We can let go of specialness and the chase to avoid death, turning toward what is truly important and meaningful. "Though the physicality of death destroys an individual, the idea of death can save him" (Yalom, 1980, p. 156).

Purpose

Living with purpose takes effort. Connection with purpose means clarifying what matters to you and committing to actions that build this path. When we extend ourselves into our values, we will run into painful emotions. Choosing to engage values involves stepping forward with our fears, moving against our own desire to hide or escape. But we will also step into joy, connecting to the here-and-now moments that animate us in the time that we have. To do this requires our attention. It requires energy. Becoming aware of our current preoccupations with thought and emotion, with the past and worry, while letting go of the struggle to control these experiences, however, invites us into possibility. What can be chosen and done as we engage the creation of life meaning? But possibility is relative only to no possibility. And no possibility arrives with death. Therefore, acceptance of death is vital to the discovery of meaning and purpose. Your personal work on this will matter. It will influence what you do in life, and it will impact what you do in therapy.

REFLECTIVE Practice 4.4

Continue to explore death. Ask yourself and take time to reflect on the following questions, being curious about what you notice:

- Have I ever been able to talk with my friends and family about death deeply?
- Have I ever had conversations about my own death? If not, why not?
- Do I ever consider the death of those around me who I care about?
- Have I ever considered the broader death experience: we all die? Many have come before, and many will come after?
- Have I ever explored my own sense of specialness? Honestly, deeply explored how I might be special or different from others? What do I notice when I consider this specialness? What do I notice when I recognize that I am not special?
- What relationship with death would I be willing to consider as I move forward? How will I work with death in therapy?

The Heart of Now and the Door to Next

Facing your own death can carry with it a weight, a sense of sadness. As Brian Jacques (2011) states, "Don't be ashamed to weep; 'tis right to grieve" (p. 59). But grieve well and be aware. And then, in the next turn of time, step forward, engage. Choose and act, taking responsibility for what happens next, for authoring your life. Whatever the choice, it is not a should or a must. There are all kinds of ways to lead a life. But choose your life and assist your clients in becoming aware in the service of choosing theirs as well. Time moves on—it is a terrible and beautiful thing to understand. How will you spend yours?

As you consider how you will author your life in this moment, in chapter 5, I will invite you to consider how you will author your therapy. This authorship will be specific to use of language itself, both as it is spoken and as it is conveyed through body language. How you speak and what you notice and convey in body language are part of the ACT therapeutic stance. I encourage you to explore your therapeutic fluency next, as you continue to author your ACT work.

CHAPTER 5

Growing Your Therapeutic Fluency: What You Say and How You Say It

It's not what you look at that matters, it's what you see.

—Henry David Thoreau

You've embarked upon your personal work on being open, aware, and engaged and have begun considering how this practice influences your ACT work. Now I invite you to examine your "stance" in therapy, as well as languaging between you and the client. Just as you author your life and are assisted by knowing the inevitability of your death, you can author your stance. I will focus more specifically on the stance in therapy in chapter 10, as stance is part of the ongoing and overarching process in ACT, and process is the focus of part 2 in this book. In this chapter, however, we will begin this work by focusing on building your therapeutic fluency, or personal way of languaging, including not only literal language but body language as well. As in previous chapters, I will encourage you to participate in reflective practices designed to help you consider your own languaging in therapy and possibilities for change if needed. As you read, think more broadly about how body and spoken language support or perhaps hinder being open, aware, and engaged.

The Unique "Languaging" of ACT

Many therapists have noted that there seems to be a kind of ACT-ease, so to speak. I agree; language or languaging in ACT tends to be used unusually or distinctively. I believe this is the case for at least two reasons. The first is about control. Languaging, or communication, carries a substantially high amount of interaction, both with the self and other, related to control. Think about how this works in your own life and you will get a sense of what I mean. Consider all the ways wherein you try to regulate, limit, constrain, direct, check, influence, command, sway, organize, contain, monitor, oversee, dictate, stop, impede, withhold, inhibit, and so on, the many aspects of your life. There is layer upon layer of verbal (and nonverbal) control, and there is layer upon layer of rule following, or behavior that is governed by rules. ACT therapists need to successfully penetrate these layers, winding their way through to letting go, defusing from mind, and ultimately speaking, with both literal words and body language, in a fashion that is directly in support of openness and acceptance.

We can start by taking a closer look at the familiar layers of control in our lives, helping us to see the difficulty we are up against. The regular workday is a perfect example. Your day often starts with control. The alarm clock awakens you. This human-made device controls the amount of sleep you get. You have a morning routine that is "ironed" out so that you can drink your coffee, have your breakfast, get the kids ready (if you have them—and if you do, take a look at all the control that goes into simply getting them ready to get out the door in the morning), and get yourself ready—shower, clothes, deodorant, perhaps makeup, cologne, perfume, shave, aftershave, facial cream, shoes (notice how we control our image). Get in the vehicle, control where it goes. Try to control traffic—this can run the gamut from speeding to honking to tailgating to cursing to following the law—and arrive at work. It's only 8:00 a.m., and you have already encountered a tsunami of control—and this is the short description. It would take hundreds of pages to lay out the language-based control that happens in a single day (it is also important to briefly acknowledge that the many behaviors that occur in a day have a multitude of varied and complex functions that are not necessarily about control). Given the massive amount of time (largely blindly) we spend in efforts to control, the work needed to access the experiential self—creating a small rupture in languaging—is challenging and, as a part of the process, perhaps means speaking in a particular way. It is shifting from control-based language, whenever possible, to acceptance-based language. "Speaking in a particular way" includes body language and posture and the embodied therapeutic stance as well.

The second reason that ACT therapists seem to have their own way of speaking is related to the function of language. ACT therapists may say unusual things based on the function of a behavior, rather than its form. That is, they may say something unexpected or completely unrelated to what the client is reporting to address the function, changing the context of the interaction in the service of behavior change. What is said is based on the therapist's conceptualization of the client and their problems from a functional standpoint, the client's interpersonal functioning, and the goals of therapy. Additionally, at least a portion of the language will be connected to its functions in relation to the broader purposes of ACT.

What You Say Matters

The language you choose to use in ACT, including literal language and body language, makes a difference. Shifting your language such that it reflects acceptance of *internal experience*, under conditional circumstances (i.e., there may be times when control of internal experience is functionally appropriate), while using language that reflects control of *behavior* is a bit of a balancing act. Two keys to making this shift involve speaking to function and speaking to consciousness. Let's break these down.

Speaking to Function

Speaking to function rather than form means being able to identify the function of a behavior and responding accordingly and rapidly. This will look different from responding to the content of what someone has said.

Let's use the language of control as an example. As mentioned, control can be expressed in language in countless ways: everything from the words "should" and "must," to straightforward rule following, to hidden attempts to manipulate, to plain-old everyday language. Also, with our posture, body language, and embodied therapeutic stance we can evoke and reflect control. But to help get us thinking about the ubiquity of the problem, let's examine the issue of control through a single word: "belief." This word tends to carry a fair bit of weight in our and the client's understanding of the world. Notice, as well, that the words "knowledge," "evidence," "facts," and "belief" are often conflated (derived) to mean the same thing. This is easily understood from an RFT perspective and is connected to a sense that something is true; there is a quality of confidence about whatever it is that is being believed. This kind of verbal "confidence" is related to other concepts such as positive, sure, convinced, definite, and, most problematic at times, truth. From an RFT perspective, there is no limitation to what belief can be linked to. Our literal relationship with this word can lock us into implied correctness about our knowledge of ourselves, our behavior, and the world. The shifting of language such that it disrupts this "truth" (speaks to function not form) is part of this different way of speaking. A short example might be useful:

Client: (*stuck, delaying action*) I have really been wanting to let go and just let my husband do his thing. I do mine. I want to stop being so suspicious, stop checking his texts and email. I just don't believe I can do it. I don't think I can stop myself.

Therapist: Belief is not required.

Here the therapist has said an unusual thing. This is not a typical way of responding to a comment about not believing in one's own capacity. It at once interrupts the kind of mind-y collapsed place that clients get into, defusing from the "sense" that belief is needed to behave a particular way and indirectly speaking to the function of the word in the client's life (i.e., fusion with the belief functions to keep the client from acting). So, when clients state that they believe that they are worthless or incapable, I work to answer with languaging that undermines belief as a whole. It isn't a matter of truth versus falsity. It is a matter of seeing the mind—seeing thinking and seeing believing as a part of thinking.

This "way of speaking" in ACT is broad in nature, meaning that it is largely based on disrupting or undermining the language of control; there isn't a specific technique to be implemented. It might include speaking a truth linked to values and committed action (e.g., the client says, "I feel like I have been wasting my time on this problem. I mean, time is running out," and the therapist simply responds, "Yes, it is"). It can be found in irreverence, perhaps invoking defusion or choice (e.g., the client says, "I should just kill myself" and the therapist responds, "It is one option"). Or it can be brought to bear through creative hopelessness in confronting resistance (e.g., the client says with some sarcasm, "What if my value *is* to suffer?" and the therapist responds, "Well, then it seems you are living your value"). These kinds of responses are designed to "rattle" the cages of individuals trapped in language, increasing curiosity concerning choice, values, mind, and so on. This way of speaking is unexpected and shifts the "verbal ground" the client is standing on; it shakes up control. This can often create a thoughtful or reflective pause in therapy: space where the opportunity for something different to happen is possible. Sometimes this way of speaking may be just enough of a tempered jolt to assist movement forward. (It should be mentioned that the examples provided are limited in their full quality as they are taken out of context. Although they are actual examples from my own practice, they should be read with the understanding that they were delivered inside a context of care, humility, and recognition of suffering.)

Working to respond to function rather than content is part of mastery in ACT. Additionally, it is essential to pay attention to another issue at play. When we fuse with mind, we can lose the sense that knowing is viewed only from our own perspective. We collapse into our personal way of understanding the world. In this space, our directly experienced knowledge of the world is shrunk. We lose the awareness that even our understanding of others can be viewed only from our own perspective, and that perspective is based on our learning history and elicited by our current context. We are limited inside of this fusion and can begin to work in therapy inside of a position of the knower. This can be subtle and lead us into ways of speaking in treatment that are less flexible (e.g., relying on explanation, convincing the other, attachment to metaphor and exercises).

Ongoing awareness of experiential knowledge and a sense of self as a larger process can assist you, as an ACT therapist, in considering ways of speaking that invite flexibility. When you work from a place where you routinely consider conscious awareness, and the knowledge and quality of perspective taking is embedded in the ACT work you do, explanation can be tempered or replaced with curiosity, exploration, and discovery. Indeed, recognizing this broader sense of self and the way in which language can falsely give us a sense of truth might even lead us to a place where there is no absolute or correct way of understanding the world in the clinical setting. Speaking from this position is more tentative, more flexible. The truth or falsity of a word, thought, or concept, and its relation to reality, is held lightly and communicated as such. We are more open to questioning, wondering, and contemplating. We are more able to see ourselves, therapy, and life as process and more likely to let go of outcome.

Speaking to Consciousness

Contacting consciousness is beyond the conceptualization of it, as any conceptualization of consciousness is not consciousness itself. Instead, this place is encountered experientially, typically a glimpse at a time, because in the exact moment you think, *This is it*, you are again thinking, you are inside of mind. To speak about or explain the observing or conscious self remains a challenge. In *The Miracle of Mindfulness*, Thich Nhat Hanh (1976) provides a metaphor to assist with the recognition of this challenge: "The mind is like a monkey swinging from branch to branch through a forest…in order to not lose sight of the monkey we must watch it constantly, even be one with it… [Observer] contemplating mind is like an object and its shadow—the object cannot shake the shadow" (p. 41). The monkey and its shadow travel together. Consciousness is the shadow to the ever-present mind. We can set the context to create glimpses of consciousness, touching it from an experiential place. ACT maintains that it is here we learn that internally experienced events (thoughts, emotions, sensations, and memories) are not dangerous; we learn they are ephemeral and natural. We observe the ongoing flow of experience. This learning needs an embodied vehicle. It is through this embodied consciousness that we have a portal to healing. One of the ways we as therapists will want to speak then is about eliciting this more experiential or felt sense, speaking to and being able to communicate with our clients about this unattached space. It is from this place, whatever the client brings to session, that it can be encountered, and the client held as whole. Simple discourse with full explanation and description and pure focus on content won't get you there—what you say and don't say matters.

When we are collapsed into our minds, our understanding of ourselves and the world is indistinguishable from what is truly present in the moment. Our experience is obscured through the filter of verbal knowledge, seeing the world concerning categories, judgments, forms, data, and understanding. Our ability to convey the ongoing flow of experience becomes more difficult, and we begin to relate to the client as a diagnosis or a problem to be solved—we language with them differently here. It seems wise to carefully consider our own use of language in this circumstance. I would argue that this is why we need to bring a wide latitude of inquisitiveness to mind: exposing our minds to inspection, expanding into curiosity, and regularly tracking experiences and consequences of behavior rather than simply trusting our minds and their interpretation of behavior. Rooting out avoidance through self-deception by reflecting on the experience of our conceptualized selves—ego, personal opinion, self-image, and worth—is part of this process. In observing our own programming and freeing ourselves from the ubiquity of our native processes of verbally based control, we open to personal acceptance and self-compassion, which is not only a guide to living ACT but also a way to embody it and to speak *from* this embodiment. To be clear, observing mind isn't an observation outside and independent of the observer; there is no objective or separate existence. It is all embodied. So, it is here that how you say it matters. Pace, tone, and body language fully participate in your languaging.

REFLECTIVE Practice 5.1

Consider your own attachments (e.g., conceptualized self, ego, personal opinion, self-image, worth). You can pick one of these attachments to focus on or several. Take time to notice how these attachments play out in your life. Notice their influence and the contexts in which they reveal themselves most. Notice the emotions, thoughts, and sensations that come along with these attachments. Take time to explore what seems to drive any clinging to each one you have identified. Next, consider:

- How might these attachments show up in the therapy room? (Think of times when your "buttons get pushed.")

- How might therapy look different, regarding what I say and do, if these attachments are held lightly?

Explore your attachments and discoveries in supervision or with a colleague. Consider whether fear is playing a role with respect to any level of clinging. Open up to that fear and notice how you might speak differently if you do. If fear is not in charge, what are you free to say?

How You Say It Matters

It is challenging to precisely describe the process of lining up with acceptance-based language rather than control-based language. How you communicate through your tone and pace will be a part of this work. Each facilitates or plays a role in experiential work and can assist in emphasizing process *as* outcome. Notice how a therapy session might change if you were to be mindful of your work, speaking more slowly, reverently, or evenly, without rushing toward the future or interacting with a client regarding tasks to be completed. A steady presence might be conveyed. Let's explore the possibilities.

Tone

Tone, in general, and in psychotherapy, represents several aspects of communication, including pitch, volume, and formal versus informal speech. The quality and strength of your voice during therapy can signify your emotion and perhaps even your intent. Speaking softly or gently may communicate kindness; it may imply the need for quiet; it may provide a paradox when thoughts seem loud and bossy, creating a sense of stillness in the midst of overwhelming emotions. Your pitch, volume, and formal and informal speech can all be used as metaphor and a way to assist clients in defusing. With all of these forms of tone, though, it is essential to remain respectful of the client.

It is also worth noting that exploring the *how* of any language shift can be mistaken for a set of rules around languaging. What I am trying to communicate here is subtle, and no specific rule should be followed with any kind of rigidity. For instance, some might confuse acceptance-speak with using a soft voice (acceptance equals soft). Acceptance doesn't equal soft. But it also doesn't equal not soft. The meaning of a soft tone of voice depends on many factors. Use of speaking softly depends. From time to time, I have listened to a therapist–client exchange on audiotape where it sounds like the therapist is trying to capture acceptance-based languaging by speaking softly or by quietly saying "uh-huh" and adding a little groan-like sound or an "aww" while the client is speaking. This might be a form of sympathizing with the client but shouldn't be mistaken for acceptance. Indeed, uttering these kinds of sounds during therapy can, in some cases, sound fake or insincere and perhaps even reinforce problematic behavior. Rather, the issue I am trying to target has more of an experiential flavor and is linked to the therapeutic stance, not a rule. Indeed, Cox (2015) quoted Jennifer Pardo, a speech communication and phonetics researcher, who said there is no "particular acoustic element which reliably determines how the majority of people feel about a voice, the closest is speaking rate, followed by intonation but it's not one thing on its own" (para 9). Pace and tone broadcast from a respectful position are humble in nature and generally deferent—the client always has choice—even if the words accompanying the pace and tone are confronting, irreverent, or humorous.

Research demonstrates that varying tone can impact how you are perceived, and that this can vary by gender. For instance, positive voices sound more trustworthy than negative voices (Schirmer, Feng, Sen, & Penney, 2019) and different intonations predict trustworthiness differently in males and females (Cox, 2015; Tannen, Hamilton, & Schiffrin, 2015). Varying your tone in the session can serve many purposes. Indeed, a flat or steady low tone can imply disinterest. Raising the tone of your voice can be helpful to emphasize an issue. Speaking quickly is usually accompanied by a higher tone and conveys a sense that you are in a hurry. By contrast, speaking more slowly is often accompanied by a lower tone. Adding pauses to a slower, lower-toned voice lets your listener keep up with you and conveys the message that you want to be there talking to them. Simply being aware of tone can help you to slow down. Finally, articulating clearly is always important, as mumbling is challenging for anyone, let alone a client. Importantly, tone can convey and evoke emotion, an essential part of psychotherapy.

Manuela's Reflection

Trying too hard to control your tone of voice may sound fake. You may find it more useful to consider tone as it relates to your overall presence. To cultivate tone, you need to cultivate your own presence—an embodied and experiential presence that is essential to ACT. From the cultivation of this presence the right tone can emerge. As well, tone can create an atmosphere that can be used functionally to evoke a more flexible repertoire in clients.

Pace

Probably the most straightforward and perhaps the wisest piece of advice I could give about pace is to slow down and breathe. Your tempo will play a role in the way you communicate with your client and will impact the way communication unfolds in therapy more broadly. Therapists can sometimes forget that style and pace of speaking have an interpersonal impact. For instance, people who speak more slowly are often seen by others as kinder and friendlier, while speaking more quickly is associated with competence (Cox, 2015). But according to Pardo (in Cox, 2015), "there's a certain sweet spot to it, if you speak too fast then you sound nervous" (para 8).

Speaking at length or in long, complex sentences can be problematic. Your client may stop listening or get lost. And, if you are planning to speak more from an intrapersonal, honest place, then shorter sentences, well-timed and paced, will be beneficial. The longer you speak, the more likely it is that the words are no longer from the heart, but have moved upward and are coming from the head. (This is not to say that "coming from the head" is necessarily problematic—just different.)

Pacing also creates space, and space invites openness. I have started to conduct a two-part exercise during training wherein attendees are paired up and split into roles. One attendee plays the "therapist," and the other plays the "client." They are asked to interact as if in a regular therapy session. The client is asked to be challenging, to bring a difficult issue to the "session." In part one of the exercise, the therapist is asked to do the best work that they can. At the end of this first round (about five to ten minutes), the therapists often note that they feel a bit stymied. The therapists report working hard, engaging in problem solving or in an ACT exercise or metaphor to try and help the client move forward, but not making much progress. In part two of the exercise, I invite the therapists, without the clients' knowledge, to simply take a single breath before each time they speak. At the end of round two, I ask the clients if they notice anything different between time one and time two. The answers are striking. The clients often report that they felt better heard and understood. They note more empathy and compassion. They state the therapist seemed less anxious and more present. They report more space and openness. The clients rarely guess that the "secret" for the therapist was to take a single breath before speaking.

Introducing this breath, this small pause, is creating something—a slower pace. Slowing down in responding to the client allows the therapist time to be aware, to connect and approach what happens next with some degree of presence. This helps to create that open and curious space. The rush to solve is tamped down, process as well as the function of behavior are more easily recognized, and the possibilities for responding in a way that assists the client are more available.

Therapists can step up the pace while speaking in therapy, sometimes without even noticing it has happened. This increase in pace can be the result of problem solving, following a protocol and running out of time, feeling nervous or anxious about something a client has done or said, or even tiredness. It may be due to impatience or frustration. Or it can just be due to that ever-present and well-learned desire, wish, or hope to control. Speeding up, however, has its drawbacks. Therapists can become confused, dismissive, or overly confident, thinking that the client is hanging in there with the work when they are not. Not only does the therapist lose contact with the moment, but they are also likely to lose touch with the client. Speeding up may be used as a form of defusion or to convey an idea (e.g., worry), but regular interaction is worth a more deliberate pace.

I once worked with a therapist named John (not his real name) who had been doing ACT for about five years. Although junior at work, I wouldn't refer to him as new to therapy. He had been well trained, attending many ACT workshops and receiving supervision in the approach for a couple of years. He asked me to listen to audiotapes of his sessions. A bit alarmingly, I discovered that he was speaking about 80% of the time in the majority of his sessions. He was also speaking relatively quickly. Additionally, and I believe this was a function of how much he was talking, John was slipping into a somewhat "jargony" process. Speed and amount of languaging during ACT therapy, whether they are an attempt to get each of the exercises and metaphors into the session or whether the therapist has accidentally slipped into a teaching mode, move the therapist and client away from openness. Here, the therapy and the therapist have become too prominent. The triangle of the therapist, client, and intervention are out of balance. There is no blame in this. ACT, as well as other therapies, are susceptible to a teaching orientation. Control, as noted earlier in the chapter, is so powerful, that mastering the content and delivering it as a lecture can seem like "doing ACT." In listening to John's audiotapes, it was clear that he was excited, he loved the intervention, and he wanted to share it with his clients. But that enthusiasm, coupled with the pace of delivery, didn't make up for what was needed in connecting to the moment and to the client.

John knew the ACT material hands down; he would score high on any measure of fidelity (he was delivering metaphors and exercises pretty much as called for in any protocol). Nonetheless, he spent so much time talking during the therapy session that he would be rated on the low end of any competency scale in delivery of ACT. The clients he worked with seemed to be overwhelmed, and the most typical response by clients that I heard on audio was "uh-huh." Pace matters. John and I worked together diligently to help him find ways to slow down, consider the function of behavior, notice process in therapy, and be more responsive to the client. Breathing, letting go of doing "it" right, opening to fear and anxiety, and being willing to sit with the client's pain were all part of bringing a more measured pace to his interactions. Perhaps any of these strategies might be helpful for you, too, even if you don't operate at a quick pace.

Pace also impacts the content of speech. Maintaining a slow, steady pace helps you to avoid using jargon and to speak mindfully.

AVOIDING JARGON

Later in our consultation process, John revealed, with a bit of awkwardness, that he wanted to "look" smart, both for the client and for me in the recordings. Indeed, he might have looked smart, and I would submit that he was smart. But it didn't matter as he both literally and figuratively lost his clients. As you likely know, psychology can be filled with jargon, and ACT is no exception—defusion, self-as-context, perspective taking, frames of coordination, framing, shaping, contingency, contingency-shaped behavior, "selfing," "languaging," building a flexible sense of self, pliance, tracking, intrinsic motivation, habituation, relational, functional, and so on…and I barely touched the terms found in RFT. Can you imagine using the words "mutual and combinatorial entailment" with a client? Well, probably not. But it is worth cautioning against the use of jargon.

SPEAKING MINDFULLY

When speaking quickly, the therapist can end up "shooting from the hip" as well. In some cases, this might be just what is needed, but more often it will miss the target. Slowing down combats the therapeutic misses or missed opportunities. Mindful speaking helps the therapist to be centered, improving concentration and attention to what is needed—effective interventions in the moment linked to the function of behavior.

I worry a little bit here that many reading about the pace of communication in therapy will think it too simplistic. Of course, it makes perfect sense to slow down; it aligns with present-moment work; it supports each of the mindfulness processes on the ACT hexaflex (diagram of the six core processes; see Luoma et al., 2017, page 17); it invites more opportunities to notice overarching, interpersonal, and intrapersonal processes (see part 2); and it allows room or space for experiencing. I would only ask a couple of quick questions: When was the last time you checked in on your own pace in therapy? If you did, what did you discover? What if you slowed the pace just a tiny bit more and at opportune moments? There is always room for growth. If we are working as therapists to evoke emotional content in support of experiential work, then the pace will assist. Slowing down will support attending to the material in the session that will help set the context for a greater experiential focus. Note the distinction between the following two possibilities based on the following therapist–client exchange:

Client: I have been very upset about my daughter, and my husband is making me crazy. He is making things worse with her.

Therapist: What has been happening?

Client: She is keeping things a secret, well, I mean, she isn't telling us what is going on at school. We pay for her college, and we don't know her grades or financial situation, we don't know if she is about to get kicked out or get a scholarship. My husband got angry and compared her to her older sister, and she lost it. I started to cry. I wish I wouldn't have done that. My daughter yelled and stormed out of the room. My

	husband and I got into a fight and ended up not talking to each other for the rest of the day.
Therapist:	(*gently breaks in*) The situation seems difficult. Sounds like everyone was struggling. What did you and your husband start fighting about?
Client:	We were fighting about how he compares her to her sister. It just isn't fair. He shouldn't do that. It makes the whole situation much worse. It just doesn't seem like the thing to do when you are talking about school with your child.

Now let's change the pace:

Client:	I have been very upset about my daughter, and my husband is making me crazy. He is making things worse with her.
Therapist:	(*breaks in*) What has been happening?
Client:	She is keeping things a secret, well, I mean, she isn't telling us what is going on at school. We pay for her college, and we don't know her grades or financial situation, we don't know if she is about to get kicked out or get a scholarship. My husband got angry and compared her to her older sister, and she lost it. I started to cry. I wish I wouldn't have done that. My daughter yelled and stormed out of the room. My husband and I got into a fight and ended up not talking to each other for the rest of the day.
Therapist:	(*breaks in, gentle tone, but slightly resolute*) Can I go back to something (*waits for several seconds, slowing the pace*)…Why do you wish that you hadn't cried?
Client:	(*tears up and cries*) I cry when people get angry.

I can't say that the client's response in the second exchange would definitely happen, but taking the time to pause and "catch" those places that seem vulnerable, avoidant, and/or difficult will alter the process of therapy.

Manuela's Reflection

It is also worth noting overall that the type of language is important. As ACT therapists, we tend to use evoking language in ACT to promote experiential work. We use less informative language or inductive language. I loved the above example; it is a good demonstration of using evoking language as opposed to informative or inductive language.

REFLECTIVE Practice 5.2

I invite you to consider your personal pace and tone in therapy. This is a great time to listen to an audio-recorded session of yourself doing therapy (perhaps do this with a supervisor or colleague as well). Listen to and notice your languaging, pace, and tone. Do they have the impact you thought they might? How do you react to hearing your own voice—its tone and pace? Notice if slowing down is needed and explore the potential impact of that on therapy. Ask yourself the following questions:

- When do I tend to rush or use jargon?
- How might change of pace and tone influence my therapeutic work?
- What if I slowed the pace just a tiny bit more and at opportune moments?

Pace and tone are also synced with body language—our tone is higher and faster when we are excited, for instance. Just as we may want to grow our awareness for pace and tone and consider how it conveys acceptance or importance of something said, we will also want to be aware of our body language. It too conveys a tremendous amount of information and can be used effectively to support ACT processes.

Body Language in ACT

Being aware of your client's body language from the moment they walk into your office will nearly always be a useful part of therapy. Being aware of your body language will be important as well. Your entire body participates in languaging, either showing or hiding your emotional state, communicating in its varied ways presence and process for both you and the client. Each of you has a body language and a presence that tells the other how you are feeling. Being aware of body language, its meaning and function, is indeed an essential part of excellent communication and mastery in therapeutic interaction. From an ACT perspective, it is inseparable from the discovery of and subsequent intervention in avoidance, fusion, and loss of contact with the moment. It is present in values-based work, committed action, and self-as-context. It is entirely part of the overarching, interpersonal, and intrapersonal processes.

Manuela's Reflection

I think that body language, both yours and the clients, is very revealing. It can assist with understanding. When you are in sync with another's presence, there is a deeper communication process that often goes beyond words. You can "feel" the other's

> *experience in a fuller and whole sense. It is an embodied flexibility, one that makes space for the therapist to tune into the client. In other words, you are attuned with the client. This is neurologically based and involves mirror neurons.*
>
> *The tricky point to remember about body language is that when reading body gestures, you can't derive these as simply words. It is not that linear. It is subtler than that. Some authors, like Lakoff and Johnson (2008), suggest that all cognitions are embodied and that abstract concepts are largely metaphorical. "Reading" a specific gesture as always lined up with a specific action or verbal interpretation places the "read" at risk and takes on a more mechanistic quality. Gestures are like metaphors (Lakoff & Johnson, 1999). So from a more contextual point of view, body communication has different layers and levels. In working with body language, it is useful to take a contextualistic perspective, to see gestures (forms) related to function in context. Body language isn't only specified by form. It is a metaphor.*

Across the years I have heard many different percentages representing how much body language is a part of communication: 70%, 80%, 93%, and others. Albert Mehrabian has published several articles on human communication (Mehrabian & Wiener, 1967; Mehrabian, 2009). His findings have routinely been misinterpreted. He has become famous for the *7%–38%–55% Rule*, referring to the relative impact of words, tone of voice, and body language, respectively. Adding 38 and 55, you get the previously mentioned 93%. However, if you review what happened with the study that led to this rule (Mehrabian, 1972), you discover that this rule applied to a particular context. These numbers occur when the nonverbal and verbal channels are incongruent (see Mehrabian, 1972). It is more likely that the degree to which body language participates in communication is quite broad and ranges from very low to very high, depending on the context.

One thing is clear: body language plays a significant role in human communication and can be thoughtfully used in therapy in varied ways, both as information for you and the client and as a point of intervention. Nonverbal cues can be revealing and potentially more accurate than verbal cues, especially when considering avoidance and its many forms. A client (as well as a therapist) often talks "over the top of emotion" in therapy sessions, for instance. Targeting or speaking to a form of body language, encouraging acceptance through body language, or using metaphor via body language can all be useful. Becoming more fully aware of your client's body language, as well as your own, will assist in this process.

Consider the following questions when thinking about your client's, as well as your own, body language:

- Are repetitive gestures present? Are they fast or slow?

- Is posture upright or collapsed over?

- What is the body position? Is it grounded? Does it evoke anything? Is it constricted? Does it seem to have energy?

- Is movement relaxed or controlled?

- Where do the eyes rest and what do they move to? Is there good eye contact?

Paying attention to and reflecting on these body experiences can be quite useful in therapy. Not only does it connect you to the body, it can convey openness or control, interest or noninterest, connection and understanding, and much more. Attending to body language brings the client (and you) into the moment, deepening awareness to the here and now.

> ## Manuela's Reflection
>
> *Nonverbal cues can be more revealing than verbal cues, especially since we know that the only way to be avoidant is through language. The body is always in the present moment, so there is no space for avoidance.*

Three Cs of Body Language

There are three key areas, or three Cs, to consider when tuning into body language (see Thompson, 2011): context, clusters, and congruence.

CONTEXT

Body language may be different in the therapy room than in other *contexts*. The situation where the communication occurs matters. Let's take the example of eye rolling. If I roll my eyes at a silly joke at a party, it will most likely be different from rolling my eyes during a serious discussion with my boss. It is worth asking what the function of the eye roll in each of these situations is. It could be the same or different. Each of these eye rolls could be about annoyance; however, eye rolling at the party could also be flirting (yes, eye rolling was a form of flirting and only took on its newer meanings over the past half-century or so; see Wickman, 2013). Eye rolling could be contempt or sarcasm. Understanding its function will tell you better how to intervene. You may choose to address eye rolling differently based on whether the behavior is a bit of playful sarcasm, contempt, or annoyance, or a little bit of flirtatious behavior. Assessing context of body language can be useful across sessions and can help you to conceptualize when and where a change in behavior is needed.

CLUSTERS

As you are becoming more aware of body language and considering its role in communication and therapy, you will want to be aware of *clusters* of body language that communicate the same thing. Viewing multiple body language behaviors can bring together a fuller picture of the story, giving the body language behavior meaning as a whole. I have one client who leans forward almost on top of his knees, crosses his legs, puts his elbow on his thigh and his hand in the air with his thumb, index, and middle finger touching. His head tilts slightly, and he looks directly at me. The meaning of this body language is *interest*. But not just any interest—this is a client who listens well and is fully participating in therapy. This set of body language behaviors tells me something. It says to me that he really wants to "be with" the exchange that is about to happen. He is highly interested. If I attend to these places and notice that change is made following these interactions, then I will want to be aware of the content and process during those moments so that I might bring them into the room again or rely on them in some way.

This sort of cluster of body language can be related to a central concept of presence as well. Therapeutic presence, as defined by Geller and Porges (2014), "involves therapists using their whole self to be both fully engaged and receptively attuned in the moment, with and for the client, to promote effective therapy." (178) This means that the client and therapist produce a gestalt, embodied by different layers of presence and communication between them. Each of their clusters of body language behavior, using their whole selves to communicate, is helpful concerning both stance and metaphor. "Closing your body down" (e.g., hunching, getting smaller) to communicate avoidance and "opening your body up" (e.g., sitting up straight, putting your arms out) to communicate acceptance are examples.

CONGRUENCE

Finally, another useful nonverbal communication process to look for is congruence between the verbal and nonverbal behavior. In other words, does what the person say match their body language? Do their nonverbal expressions communicate the same meaning as their verbal expression? Does the spoken word match the tone of their voice? Do their mannerisms and gestures ring true to a specific emotion?

If the client is stating that they don't know what to do in a difficult situation while shrugging their shoulders and raising their eyebrows, then the communication is *congruent* (as well as clustered). Congruency can be particularly important for the therapist when the nonverbal behavior and verbal behavior do not match. It is good to pay attention to *when* they match, or don't match, as well. This will give the therapist information about integrity, honesty, and trust, for instance. Or it might provide the therapist information about willingness. That is, does the client breathe deeply, relax their posture, and maintain eye contact during willingness work, or do they look down and fold their arms right after agreeing to be willing? The therapist may approach these two scenarios differently—in the first, moving ahead with the willingness exercise, and in the second, meeting the client where they are at and checking to see if control is still present, pointing to the body language. You can certainly comment on body language that is congruent (e.g., "When you smile and laugh while talking about your child, I feel joy myself and feel like something really important is in the room"). When verbal and nonverbal behavior don't match, you may want to cue in. The nonverbal behavior may be a more accurate sign as to what is happening than the verbal behavior (Vrij, Edward, Roberts, & Bull, 2000). Speaking to the nonverbal behavior may be the better intervention (see example below).

How do we attend to congruency when dealing with clusters? You can attend to a cluster of nonverbal behaviors that are incongruent with a verbal report, or you might choose to home in on a single event of body language. I had a client who would smile and frequently laugh during a session. Most of this behavior was congruent and didn't need to be addressed, or we reflected upon it appropriately (e.g., values-based, interpersonal process). It wasn't causing her functional problems. Indeed, most of the time it was charming. She once told me a painful story of humiliation, however, while smiling. Her body language did not match her verbal story. At that moment, I gently asked her about the smile. She noted that she hadn't been fully aware of the smile but suspected she would cry if she told the story without a smile. I invited her to give it a try, and sure enough, she was right. She cried. And based on her extensive attempts to avoid crying, it was just what was needed in that moment of pain. Toward the end of the session, after reflecting on what had happened with the smile during the painful story, she noted that she had not wanted to cry in front of me specifically. As a fellow therapist, she thought I might think less of her. We explored this a bit from a compassionate space, both sharing our desires to be liked and respected by our colleagues. We noted the irony of smiling while in pain, while being a therapist, sitting with a therapist, doing therapy. We laughed then too.

Another way to observe congruency might be idiosyncratic, and you may know about it only after you observe the client over time. Does the client always make the same body movement when they are feeling uncomfortable (small wiggle in chair or clearing of throat) or about to shut down emotionally (looking down or furrowing their brow)? Watching for these kinds of body language movements and detecting patterns might help you in planning and making interventions, for instance by changing what happens next in the session. Rather than letting a client continue with a story that functions to avoid emotions, if she makes that same kind of shift in the chair that predicts avoidance, stop her and ask her about it. You might say, "I have noticed in the past that when you shift in the chair like that, you are about to say something difficult—something hard—but you move away from it really quickly; you run over the emotion with your words. Have you noticed that? (*wait for client response*). I invite you to speak much slower as you tell this story; let's see what shows up." This kind of intervention slows the process down and gets you and the client to be more aware and more present while providing the opportunity for exposure to difficult emotional material.

Other Body Language Considerations

A few studies have been conducted on small amounts of nonverbal behavioral information, known as "thin slices" of behavior (see Ambady & Rosenthal, 1992; Hall & Bernieri, 2001). This small body of research found that people can accurately judge others' emotions at above chance levels based on remarkably small amounts of behavioral information, sometimes in less than one second. Although incredibly rapid, it makes sense when thought of concerning survival. For instance, we need to be able to judge anger and aggression rapidly. There is also a body of research regarding the measurement of *nonverbal sensitivity* (Hall & Bernieri, 2001). Nonverbal sensitivity is your capacity to recognize the communication of emotion or intention through expressions of the face, voice, body posture, and gestures. This ability to judge another's emotional expressions is one of the defining aspects of emotional intelligence (Mayer, Salovey, Caruso, & Sitarenios, 2003).

Unfortunately, most of the research to date has not looked at the ability to convey emotional messages as intended. Being able to nonverbally and accurately express emotional intent is a useful modeling skill in therapy and is undoubtedly part of effective ACT interventions. Many therapists are modeling appropriate nonverbal expression of emotion for their clients. I have wondered about messages inside of non-ACT training programs and books that encourage the therapist to remain in a kind of unknown state, revealing nothing to their client about their emotional experience. My hope is that by being aware of the importance of interpersonal capacity to identify and convey emotion as a skill, and as part of a healthy relationship, that the stance of blank slate can be left behind and we can work to model and share with our clients both verbal and nonverbal aspects of emotion. Indeed, it has been asserted that emotional self-awareness is a prerequisite to insights into others' emotional behavior and empathic ability (Cohen-Cole, 1991; Matthews, Zeidner, & Roberts, 2004). With this in mind, I think it is essential to consider the full range of possibilities when being aware of body language and how it might be a part of the therapeutic process and change. Attending to thin slices of behavior and increasing your awareness support the growth of your nonverbal sensitivity and can assist with emotional exchange in therapy.

When a client is telling a story about their experience, perhaps something that occurred during the week between sessions or something from the past, I agree that hearing the content of the story is essential. But listening with awareness means broadening the perspective and not only hearing the words, the content, but also "hearing" the gestures. Speaking to them directly, instead of simply the content, may be quite useful in understanding what is happening. Making an intervention that is process oriented and speaks to function rather than form may be effective as well. I once supervised a therapist who had a client who started every session sitting up firmly in her chair with her arms folded, her legs crossed (the leg on the top swinging back and forth a bit wildly), and her jaw set, eyes intently focused. She was communicating something. Her chief complaint and reason for seeking therapy was her relationship with her husband, feeling as if he didn't understand her trauma history and its impact on her, feeling as if he had no compassion for what she was going through. She noted an extended history of being stuck in the relationship, each of them "dug in" on their view of the other. The therapy was also stuck; it was going nowhere. And every session started the same: sitting firm, arms crossed, legs crossed, jaw and eyes set.

The client would talk about her relationship, its arguments, its failings, and so on. The therapist was intimidated and mainly listened to the stories throughout the session. Yet there was a "beautiful" metaphor for the problem of control and a potential way to get at willingness and committed action, right before the therapist. Many thin slices of behavior sat before her. To help the therapist get unstuck, I worked with her to shift from the content of the stories to the content of the body language in the present moment. In this extended vignette, notice how the body language is weaved throughout to suggest the distinction between control and willingness.

Therapist: (*observing the client's posture*) I have noticed how you come into therapy.

Client: (*defiantly*) This is how I am.

Therapist: (*gently*) As we connect to "This is how I am," I will invite you to be aware of your body and what it is doing.

Client: (*shrugs shoulders*) What do you mean?

Therapist: Well, I notice that your foot is tapping up and down, your arms are folded across your body, your legs are crossed.

[Client gives a hefty sigh.]

Therapist: (*referring to the heaviness of the sigh*) Can you feel that?

[Client shrugs.]

Therapist: Let's see where you are now and…noticing right where you are, be aware that this is where we start every session.

Client: I thought I was supposed to be getting better; if we always start here, then this sucks.

Therapist: Inviting you to be here, noticing your body, again.

Client: (*with slight sarcasm*) Well I am sorry for my body.

Therapist: There is no request here but to notice.

Client: (*with sarcasm*) Okay.

Therapist: (*staying with the process, being aware that the sarcasm is part of the same behavioral pattern*) As you notice this across time, that every time we start here, is there something that you could do or I could do differently that might change what is happening?

Client: (*with frustration*) I am stuck, and I am pushing you away. I might as well just shoot myself right now.

Therapist: That is an option, but not the greatest. You want to build intimate relationships, you want to build the relationship with your husband, you want to feel connected and get closer to your husband.

Client: (*with sarcasm*) That one is done for.

Therapist: (*inviting the client back to possibility in the moment*) Let us come back to building. How can you and I, right now, in here, work on building in the moment?

Client: (*slightly softening*) I don't know…do you mean like…I don't know.

Therapist: (*gently*) Notice your body.

Client: Okay.

Therapist: Is there something that could be done here, an action, is there something that you could do, right now in this session, that would open it up a bit?

Client:	I could uncross my legs.
Therapist:	Would you be willing to feel what you are going to feel and uncross your legs?
Client:	(*with slight sarcasm*) So you think if I uncross my legs that will help us to have a better connection.
Therapist:	I don't know; we can only look and see.
Client:	It's hard.
Therapist:	Do you want to start with something more manageable?
Client:	I could uncross my arms.
	[Client laughs uncomfortably; therapist waits.]
Client:	I am just pinching the inside of my arm right now.
	[Client then opens her arms and places them on armrests.]
Client:	I feel unprotected.
Therapist:	Notice what you feel in your body; notice the vulnerability and at the same time notice you and I are now doing something different.
	[Therapist waits, allowing time for the client to experience the emotion and sensation.]
Therapist:	When you opened your arms and felt vulnerable, what else did you notice?
Client:	I felt way more anxious. My hands started sweating.
Therapist:	Simply notice sweating hands; be aware of your experience. See if you can stay here with me. You and me sitting here talking about sweaty palms, but it is a step closer than where we were.
Client:	(*pause*) So, what do I do about my husband? Uncrossing my arms isn't enough.
Therapist:	What if you can take this with you and do it with your husband? Time is running out.
Client:	You're right, it is…Okay, now more anxious.
Therapist:	You have between now and the time you are not here anymore, and the clock is ticking, and every session has started here, stuck…and now you have opened your arms and are being a little vulnerable…a tiny step toward something different, in the room right now honoring your value

with the clock ticking…and now you are engaging me differently… Would you be willing to do this if it means you get closer to that valued thing you want with your husband? You get to choose.

Attending to body language in this interaction was key to movement forward in the session. The client not only spoke about her avoidance, she physically expressed it. Observing thin slices of behavior and engaging nonverbal sensitivity played a role in working with the client on willingness and growing connection—the client's stated desire.

Elements of Body Language

There are numerous forms of body language, including mutual gaze and eye contact, gestures, skin tone, and spatial presence and distance. Let's take a look at these.

MUTUAL GAZE AND EYE CONTACT

Bensing, Kerssens, and van der Pasch (1995) found that physicians who frequently gazed at their patients were more successful in recognizing psychological distress. Greater eye contact resulted in more effective reading of emotional cues, supporting easier recognition of distress. Bensing argued that eye contact enhances listening skills and the ability to interpret verbal and nonverbal cues of distress more accurately.

Eye contact is a fundamental part of connection. Starting with a baby's gazing into their parent's eyes, it continues to be one of the most critical parts of human interaction right up until death. We read everything "in the eyes," from emotion, such as sadness and laughter, to age and meaning. Eye contact, or no eye contact, will tell you the level of connection you have with a client as well as what their eye contact is potentially like outside of session. I would argue that eye contact is probably one of the more important parts of body language to attend to in therapy. It will inform you about the level of intimacy and bonding between you and your client. It signifies interest, something you will want to convey to your client as well.

It will also tell you a bit about the client's avoidance strategies and level of willingness. It may "speak" to shame, guilt, anxiety, fear of evaluation, shyness, or past abuse. Watching your client's eye contact change over the course of time or within a session may be informative on many levels—perhaps indicating a shift from isolation and fear to one of being more open. Of course, there are cultural norms for eye contact, and these can vary quite a bit. Thoughtfully considering those norms will be useful when interacting with your clients. Ultimately, eye contact as a cultural norm can be considered a part of verbal behavior and "stepped" outside of. It's best to be wise about this, though, especially when considering the "Eyes On" exercise (see Walser & Westrup, 2007) done in ACT. I have not often come across a therapist with poor eye contact (although I certainly have), but I might boldly say that if this is the case for you, it will be challenging to do therapy in general, and quite challenging to do ACT specifically. Showing up to your client with heart, warmth, and compassion will require it. It is an act of willingness itself.

GESTURES

A favorite body language area of mine to pay attention to is gestures. They express ideas, opinions, and emotions. They can be small and barely detectable, like the tiniest shrug of a shoulder, or they can be substantial and demonstrable, conveying to you a world of information. Not only can gestures inform you about the intensity of emotion, for instance, but they can also carry meaning and power. You can go to Wikipedia for a neat list of all kinds of gestures (see https://en.wikipedia.org/wiki/List_of_gestures); it's fun to learn their cultural significance or origin. Gestures are metaphors, and attending to them in session is part of "listening" to body language and listening well to your client.

SKIN TONE

Many of us turn red under various circumstances. The wonderful dreadful blush. It can happen when we are embarrassed, nervous, anxious, sick, or perhaps for other reasons. Blushing is one of the most apparent signs regarding emotional experience. And many of us do not like to blush. We worry it means something about us socially. Somehow the skin turning red is a sign of weakness and possible instant destruction. Noticing when your client is turning red is essential, and depending on the context, should largely not be ignored. Now what you say will matter. I don't tell clients, "Oh, yikes, you are turning red. What's going on?" Rather I might say, "Is something happening right now?" "What are you noticing about what you are experiencing?" If they speak to the blushing, I will ask further, "Are you fighting something or wishing something wouldn't happen?" I am working to ask questions that are focused on getting at the avoidance or the function of the experience. I don't want to shame the client inadvertently. What you say matters.

SPATIAL PRESENCE AND DISTANCE

I once had a client who would stride into the room and sit in the corner of a sectional couch (the farthest point away from me) and then spread his arms wide over the back of the sofa and spread his legs apart, pushed out straight, heals on the floor, toes pointing upward. I was struck by his posture. It felt "big," so to speak. It was part of his learned style, but it reflected his need to express bravado. It protected him from getting too emotional. He was in charge, and he wasn't going to let his emotions in the room. When he finally spoke about his father kicking him out of the home at age fifteen, he took his arms off the back of the couch and held them by his side, hands in his lap. He bent his knees and pulled his feet flat to the floor. This was the first time he cried in session. It was valuable time in therapy, and it was on that day that he agreed to speak with his father about the instance, as he had never forgiven him for it and it had permanently changed his life. He came back the next session and threw his arms on the back of the sofa, legs apart, knees straight, toes up. But from then on, when he changed his posture, I paid particular attention. A potential opportunity was coming, perhaps one of emotional willingness. Ultimately, I wish I had asked him to change his posture more just to get a sense of what would happen if he did. If I could go back and redo the sessions, I would focus on this posture and self-placement in the room and their capacity to help him have control.

Recognizing where and how your client sits is a part of body language and can be used to assist willingness and other processes. You can use space to invite willingness, change the context, do perspective taking exercises, or address avoidance. Where and how *you* sit in the room will also play a role in the interpersonal process. You'll want to attend to it as well and be aware of what you are communicating.

Awareness of Your Own Body Language

Psychotherapy training typically focuses on intervention and techniques without attention to how the therapist can cultivate the state of being present to the client. As well, the science of body language is not exact. It is a complicated system and will depend on learning history and context. In establishing a strong social bond (i.e., a therapeutic relationship) that is potentially helpful or healing for a client, you will need to consider what you communicate with your body. Your presence and modeling will provide the client with an environment that is safe and allows engagement in therapy. Through present-centered relating that includes eye contact, softening and warmth of voice, emotional attunement, and in-the-moment engagement, the client perceives safety, and the invitation to openness is set. A therapist who takes an open and interested posture communicates a receptive and accepting stance, providing a context that may promote openness, awareness, and engagement.

REFLECTIVE Practice 5.3

I invite you to sit in your "therapy chair," either literally or in your imagination. Sit as you typically sit. Close your eyes and notice, take your time, as if you were doing a slow body scan. What is your posture? What movements do you make? What is being communicated through body language? Imagine a challenging client in front of you and notice if your body language changes. Look closely. Do you move your body or mostly hold still? Are there thin slices of behavior that are "tells" of your emotions? Explore your bodily presence in therapy and see if there is anything you want to change. Open a discussion about ways of speaking and body language with your supervisor or colleagues.

The Heart of Now and the Door to Next

One aspect of developing competence and mastery in ACT is communicating in such a fashion that mind is recognized for what it is and held by a being larger than itself—and assisting others in doing the same. It is about seeing the broader process of thinking at work and connecting to a sense of self-as-process as well. Using language in such fashion as to convey this knowledge will assist in your ACT work. If we can't bring a kind of curiosity to our own thinking, our personal knowledge, if we can't in some way navigate through life without observing our mind with inquisitiveness, questioning its understanding of the way we and the world are, then we lose freedom, we lose process. We become slaves to our way of thinking instead (even ACT isn't the answer). We lose connection with how we speak and what we say.

Therapy is a process, however. It is a series of events unfolding between two people who influence each other in exquisitely subtle ways. A breath, a sigh, or an averting of the eyes can communicate more than a long stream of words, especially when therapists or clients are most vulnerable or emotional. And because this process of mutual influence is often subtle, fast, automatic, and ephemeral—or so extended over time that it's hard to connect the dots—it's easy to miss what's really happening in the moment. It's easy to lose that therapy occurs in the here and now and unfolds in the ongoing sequence of present moments in an embodied way. It's also easy to overlook the tiny, tentative moments of new possibility that can be detected through body language in which more profound change might take root. Growth in awareness to the way you speak—bringing acceptance and connection to the conscious being and embodied presence, what you communicate through your nonverbal expressions—will water the seed.

We now turn to part 2: "Building Heart Through Unfolding Experience." We will explore overarching, interpersonal, and intrapersonal process as well as the stance of the ACT therapist, further guiding you to connecting with the heart of ACT.

PART 2

Building Heart Through Unfolding Experience

CHAPTER 6

Engagement in Process

The good life is a process, not a state of being. It is a direction not a destination.

—Carl Rogers

One of the messages I have begun to focus on more directly and in more varied ways when training others in ACT is about *process*. The message, as noted in part 1 of this book, is "There is no arrival." This is intended to have multiple layers of meaning and speaks to what is intended by a series of changes and adaptations that are made in therapy that are ongoing and driven by the function of behavior—or, said otherwise, the *process* of ACT. There is no arrival in completing ACT training, no arrival in any of the six core processes, and no arrival in psychological flexibility. There is no place to stand that is permanent and doesn't require the acknowledgment of fluid experiencing and change. Thus, ACT not only has six core processes that furnish the three pillars of ACT—open, aware, and engaged—ACT is a process itself, in both its implementation and its lived quality by client and therapist. Your personal journey with ACT, as well as the client's, involves a *systematic and ongoing engagement* in mindful awareness and committed action, working together to achieve a higher purpose.

In part 1 we explored more deeply your personal engagement in the three pillars of ACT, as well as death and existence and their ties to values. We also explored therapeutic fluency as it pertains to what and how you communicate in therapy. In this and the next three chapters, we will focus on growing your recognition of and engagement in ACT as a process specific to these three processes: overarching and ongoing, interpersonal, and intrapersonal. I will briefly introduce them in this chapter and then explore them more fully in the chapters that follow. Also, although no chapter is dedicated to the six individual core processes, we will examine them more generally in what lies ahead. We will also take a brief but closer look at content in session to mark its importance. In the last chapter of part 2, we will focus more fully on the therapeutic stance and how, in its fullness and with the integration of all levels of process explored herein, it brings a certain kind of "presence" to the ACT therapeutic work.

Processes in ACT

"Process" may be defined as a continuous interaction or series of changes going forward or on. In ACT we are focusing on the function of behavior. We can define "function" as the purpose of a behavior (action) based on an individual's learning history and current context. Holding these two definitions, the process work done inside of an ACT session is about conceptualizing and influencing the ongoing behavior in the context of the client and therapist and the relationship between them in a dynamic interaction, across time. And this dynamic interaction is in the service of psychological and behavioral flexibility. With this in mind, I will suggest that there are at least four levels of process in ACT: (1) an overarching ongoing process of therapy—the arc of therapy and growth and change in direction over time; (2) the interpersonal process—the dynamic interaction between therapist and client involving emotion, languaging (verbal and body), and behavior in the service of connection and influence; (3) the intrapersonal process—the awareness to the fluid experience within the self (and when responded to, its potential effect on others); and (4) the six core ACT processes. The latter, interwoven among the first three, create the systematic and ongoing engagement of client and therapist in a larger endeavor designed to create and instantiate psychological flexibility. This endeavor—along with the therapist's personal work in being open, aware, and engaged in the service of their therapeutic stance and the ongoing dynamic flow called process—supports the heart of ACT.

Overarching and Ongoing Process

The overarching, ongoing process of therapy begins when you first meet the client and ends on termination (in a literal sense, the experience of your work in therapy will continue on with you as you move forward in other sessions). This overarching process is linked to case conceptualization as well as the continual evolution of the therapeutic relationship informed by the balancing of "presence" of the therapist, presence of the client, and presence of the intervention over time (an issue we explored in chapter 1).

CASE CONCEPTUALIZATION AS ONGOING PROCESS ACROSS TIME

There is a detailed and informative chapter on case conceptualization in *Learning ACT* (2nd ed.; Luoma et al., 2017). Understanding the work done in ACT case conceptualization takes time, and a good part of the effort will be about increasing your verbal knowledge or understanding of the material. That is, being able to formulate cases well partly involves reading and grappling with the fundamentals of ACT regarding its philosophical premises and underlying theory.

Another aspect that informs and guides case conceptualization is coherence to the model. Coherence is about approaching your work with the client from a unified whole, a consistent and logical perspective. ACT is based on principles of behavior and has its philosophical roots in *functional contextualism* (i.e., the purpose of a behavior inseparable from context, both historical and current; see Gifford & Hayes, 1999, or Hayes, Hayes, & Reese, 1988, for a broader exploration). In conceptualizing a case, working to understand the contingencies that shape behavior in a given and historical context guides the clinician in what behaviors to target and influence as related to a particular goal (i.e., values-based living). Coherence, as it relates to the therapeutic process and plan for intervention, affords logical organization and an overall sense of understandability. It continually guides what is happening in therapy, providing a foundation for exploration of the current context, the behavior of the therapist and client, and the interaction of the two across time.

One question to ask yourself as you consider your philosophical approach to understanding human behavior is *How do I think human beings work?* Reflecting on this question will assist coherence, or the logical organization of how you view a client's behavior, thus guiding you in knowing what to do in therapy. For instance, do you hold that humans are like machines (e.g., minds are like computers), and when they suffer it is because the machine is broken? (I explore the example of mechanism to demonstrate the utility of the question. I also briefly explore contextualism as it concerns how human beings work. However, it should be noted that there are other philosophical approaches pertaining to understanding human behavior; see Hayes et al., 1988.) If yes, then a coherent plan or case conceptualization involves therapeutic interventions that are about rewriting "malfunctioning code." Metaphorically speaking, the work in therapy will be about fixing the ones and zeros that are causing the problem. In conceptualizing the case, then, you might be asking what the negative and irrational thoughts are. What emotions are these thoughts causing? What kind of black-and-white thinking, overgeneralization, or catastrophizing is happening? Working to find core beliefs that are dysfunctional and restructuring (i.e., fixing) them is an example of this approach. The interventions chosen should consistently adhere to this model, supporting a coherent intervention across time. The client's behavior is linked to their cognition, and ongoing intervention targets the same.

Or, do you hold that human beings work as whole organisms, and human behavior is an ongoing act in context (i.e., contextualism)? From this perspective, human behavior isn't caused by smaller parts working to support a larger whole; it is about the whole person in a particular context (with a specific learning history), and the function, or purpose, of the behavior is the given target for intervention given that context. If you hold the assumption that human beings work this way—that they are responding in this situation based on a learning history and that their behavior serves a purpose, or function (e.g., escape or avoidance from aversive experience, excessive rule following)—then your overall approach will be guided by how to change the context in the service of influencing behavior (e.g., reducing avoidance). Conceptualization will include the ongoing act in context. Here you will be asking, over time, in any individual session, in any moment in session, and across sessions, "What is the function of this individual's behavior? What are the patterns of behavior and what is their purpose? And in which contexts are they elicited? When and where is this person engaging in behavior that is insensitive to the context, and how is that behavior working for the client more broadly?" These kinds of questions guide the coherent planning and organization regarding the case and, ultimately, the interventions. It is the underlying philosophical assumption held in ACT.

If the question *How do I think human beings work?* is never considered, and therefore there isn't an organizing set of principles by which to conceptualize cases, or on which the therapist bases their interventions, then a smattering of interventions will do. Tools and techniques become important—they *are* the organizing "principles." They are done for their own sake or because they are in a therapy manual or because they were demonstrated in a workshop, and the why and when of doing them is lost. A therapist learning and growing in their understanding of ACT who "misses" the organizing principles, for instance, might begin to do exercises based solely on the content of the client's presentation, rather than the function of the current behavior or pattern of behaviors.

If organizing principles are helpful, *if* they assist you in knowing what to do, then a good question to ask is: *How do I think human beings work?* The invitation is to engage this question fully. Dig into the philosophical and theoretical assumptions underlying ACT and see how they can guide your approach to your therapeutic work. In short, coherence is the scaffolding for conceptualizing the case across time. It is always in play as you continually assess and reassess the client's behavior and its function over the course of therapy.

REFLECTIVE Practice 6.1

(As a reminder, you can download these Reflective Practice exercises at http://www.newharbinger.com/40392.) Consider your training in becoming a therapist. Reflect on the organizing principles you were taught in your training. Explore the following questions:

- What is/was my theoretical orientation and how does/did it inform my organization of an intervention? How does/did it tell me human beings work?

- Was my original training different from what is espoused in contextualism (the function of behavior in context)? If so, how?

- (If you are interested in ACT and have a different training background): Have I considered the differences and how they might impact my case conceptualization? How might these differences influence the overarching and ongoing process?

- How do *I* think human beings work? What causes psychological struggle?

- Do I feel and demonstrate coherence in my approach to suffering?

Consider how the answers to these questions might influence your work with clients. Notice if any shift is needed to enhance your ability and skill with ACT.

THE CONTINUAL EVOLUTION OF THE THERAPEUTIC RELATIONSHIP

Another aspect of the overarching and ongoing process of ACT therapy is the continual evolution of the therapeutic relationship. This evolution may involve strengthening of the bond between client and therapist, and it may involve a break in the alliance. It may include a range of emotional reactions between client and therapist that lead to supportive and thoughtful work as well as interactions that lead to the dissolution of the therapy. Recognizing how the therapeutic relationship is developing across time is informative for the flow of the therapy. A common complaint in the field of psychology concerns learning and using evidence-based therapy protocols in session (Aarons, 2004); the evidence-based models don't necessarily capture the richness of the client and therapist relationships (Margison, 2001). There is a broader literature and significant work being done in this area, but this complaint captures part of the problem associated with attachment to techniques and protocols, as we've discussed. Of course, knowing techniques and protocols is essential; it can help to organize the overall "story" of ACT for the therapist as well as delineate the intricacies of exercises and metaphors. And some protocols speak to the therapeutic relationship, albeit this is often a briefer mention of what the relationship should entail (e.g., empathy, compassion, genuineness); or a chapter is provided on these qualities, but it isn't always explored as part of the case conceptualization and other ACT processes.

The evolution of the relationship across time is challenging to explore in detail as it can take any number of twists and turns. However, being mindful of this complexity is important. Awareness of the kind and quality of the interactions and of the function of the client's and therapist's behavior within the interpersonal interaction is part of mastering ACT.

For instance, there may be moments in the therapy where what the therapist thought to do next is not going to work based on what just happened interpersonally. Imagine that the therapist has formulated that a client is struggling at work because he does not like to be told how to do things and his boss is telling him to do his job a particular way. It is an issue of control. The client feels disrespected and therefore rejected when others critique the way he does his work. The client becomes defensive to avoid feeling rejected when others suggest alternatives. Based on this conceptualization, the therapist is planning a willingness exercise specific to these emotions. Just as the therapist begins the exercise, the client quickly shifts, gets defensive, and interrupts, asking if the therapist is trying to control, trying to tell him how to do things. In this moment, the therapist has a few options: press on with the exercise, ask for clarification, stop working on willingness, or confront the client. What to do next depends. It will depend on the quality of the relationship over time as well as the function of the current behavior in this context. Pressing on with the exercise might be a perfectly viable way to proceed if the relationship has evolved in such a fashion that this would not harm the current interpersonal process or the ongoing relationship. Asking for clarification or confronting the client might also be perfectly fine alternatives. The answer as to what to do next is, again, it depends. Remaining sensitive to the quality of the relationship, body language, learning history, and how these behaviors are functioning will be a part of the evolution of the therapeutic relationship across time. It will be part of a coherent case conceptualization in the ongoing and overarching process of ACT. This will require personal flexibility and awareness.

Interpersonal Process

The interpersonal process is about the exchange between therapist and client. This is of central importance to the unique, engaged, and dynamic quality of the relationship between two people working together in therapy in efforts designed to promote change, growth, evolution, or whatever movement is desired and agreed upon by the two. In this vein, ACT supports a reciprocal influence process that *includes* the therapist's behavior and learning history as it unfolds in the context of therapy. The therapist is a full member of the exchange rather than just a blank slate or technician applying techniques to a client. Awareness of the quality of the relationship, the impact of your behavior on the client and vice versa, and the participation of your intrapersonal process (a topic we turn to in a moment) are all part of this exchange. The interpersonal process is also supported by your chosen stance concerning a consistency in responding to client thoughts, feelings, and sensations.

Being aware of the interpersonal process might include commenting on a larger pattern of behavior between you and the client in session. For instance, if a client keeps returning to a story that you have already explored in terms of its function and emotional impact, then you might "point" to the *return*, not the story itself, and be curious about how it is working in terms of therapeutic progress, noting its impact on your relationship. You might then invite the client, as well as yourself, to track this pattern, noticing its function in the therapy session and what the hope is behind returning to the story (e.g., feeling understood, solving a history that can't be solved, needing to feel heard, avoiding other work in the session). Bringing a kind of curiosity to the more extensive process can be useful.

The kind of emotional intimacy you share with the client is another aspect of interpersonal process to attend to. Does work between you entail vulnerability? Do you feel connected to or distant from the client and does this change across time or even in a single session? Do you have feelings of affinity and caring for your client? Do they return these feelings? How you and your client relate to each other can impact motivation and expectations as well as your communication and ability to collaborate. If you don't attend to the interpersonal process or you get frustrated by it, you may begin to engage in behaviors that lessen your influence (e.g., continually checking the time, yawning, fidgeting, getting caught in your head). Cultivating ongoing awareness of the exchange between you and your client and observing how you shape each other's behavior will assist in informing your ongoing process of conceptualizing the case in the service of desired outcomes. More importantly, your stance in therapy and toward your client will be played out in the interpersonal process. Your compassion (being able to sit with them in pain), wisdom (delivering authentic messages in the service of change, or in other words, consequences), and "pragma" (or enduring love for your client which transcends the casual and is exhibited in patience for and acceptance of other) will set the context for change.

Intrapersonal Process

Finally, there is an intrapersonal process in ACT psychotherapy. "Intra" literally means within. The intrapersonal process is about what is happening and flowing within yourself—the comings and goings of your internal experience. Being awake, alive, and curious about your internal states—your senses, your emotions, your mind—as part of a larger practice is the foundation of self-knowledge. Self-knowledge, as explored in part 1, helps us to know which of our stories and emotions capture and blind or paralyze us in the therapeutic session, and in what ways. It is also important to know by which means they can serve us, guiding our actions in therapy in functionally appropriate ways linked to the desired outcomes.

Our learning history, stories, and associated emotions will be fully part of the therapeutic experience. If the ACT core competencies suggest that we be genuine, authentic, and open; that we self-disclose in the service of the client; and that we shift in therapy in flexible ways that accommodate the process in the moment, then it follows that knowing your honest experience and where and how you react will be part of what it means to engage ACT more fully and skillfully. Therapists' emotional experiences, as well as their values, influence the therapy. The notion that the therapist should remain neutral is a direct contradiction to genuine and authentic interaction. I am not saying here that all that a therapist thinks and feels is fodder to explore in therapy, but it is fodder to consider.

Your emotional reactions during the session are essential pieces of information that may assist you in understanding what the client elicits in the world outside of therapy; they may help guide you in explorations of interpersonal impact. For example, if the client is evoking your frustration, pity, anger, or boredom, this will be a useful piece of information (as we will explore in chapter 9). Skill in ACT is about being aware of these reactions and responding to them in ways that are workable, moving the client closer to their valued directions. The key is knowing how to make the distinction between when to respond and when not to respond (instead gently observing and choosing an alternative response). It is the rushed therapist—the one clinging to a favored technique, the one who has one or two standard "ACT moves," the one who is clever or rises above the client, or the one who avoids in moments of intimacy—who misses the sometimes subtle distinctions, instead running over the top of the client, perhaps providing an oversimplified or off-target intervention, maybe even avoiding what is there to do next. Slowing down and showing up to one's own experience are perhaps the best antidote. Taking more time and being consciously aware are the beginning of increasing one's ability to make these distinctions. Having awareness is the first step in behavioral choice. Knowing whether your reactions are old learning histories bumping up against the client's presentation and may not call for action, or whether your emotional responses are important feedback to the client at any moment in time, only comes through awareness to and understanding of these intrapersonal experiences.

It is the same kind of thing we ask clients to do—to be aware of their experience. In the context of intrapersonal process for the therapist, then, the question asked earlier—"Do I have to practice mindful awareness to do ACT and its mindfulness work?"—answers itself. One *can* implement ACT without personally engaging in this process. I will submit, however, that the therapy will look, feel, and work differently than it would if the therapist practices mindful awareness. Indeed, the former will more likely take on the character of rushed, technique-oriented, and potentially avoidant qualities. There is a pearl of wisdom in cultivating the practice of taking more time and being consciously aware. It assists us in moving beyond limited or fixed views of ourselves and the way therapy should be. It allows us to "see" the push and pull of our own minds, our reactiveness to it, and our personal stories and emotions and how they play out in our lives—and in our therapy. We can work to understand the function of our behavior and assess how it is functioning in a therapy session, helping us to know if it will be useful to the client or not. Recognizing how your emotional response might influence the client's behavior, should it be revealed, is part of using the intrapersonal process in therapy (e.g., does it bring awareness to something, reinforce a client's behavior, or invite curiosity?)—as well as the interpersonal and ongoing and overarching processes of therapy.

> ## Manuela's Reflection
>
> *Working on overarching, inter-, and intrapersonal processes sounds like a lot! Learning to engage in each of these processes might feel discouraging. How do you know where to start? It's like using many balls when first learning to juggle. As with any learning experience, it takes slowing down, patience to practice—even when you think you are not getting it—and the courage to try again when all the balls fall. And remember, there is no arrival…*

REFLECTIVE Practice 6.2

I invite you to review your ACT clinical work gently. Take time to connect to how you are currently implementing ACT. As you explore your work, ask yourself the following questions:

- Do I find myself overusing a technique (e.g., using it because I like it versus using it functionally)? Are there places in my ACT work that feel stilted or old? Are there ways in which considering the overarching and ongoing process might be useful in breaking me out of any stuck places?

- Do I spend time on the interpersonal process in therapy? Do I consider my stance concerning the client? How do my clients shape my behavior? Am I aware of my impact on the client?

- Do I tend to shy away from the intrapersonal elements of therapy, and am I demonstrating any kind of avoidance in that process? How can I challenge myself to grow in this area?

Simply explore and, if you're willing, commit to the discomfort of change. Where do you think you have a need for personal growth? Do you have any resistance to looking closely at these processes and how you are implementing ACT?

Slipping into Content, Staying in Process

In listening to therapists' audio recordings of ACT sessions following workshops, one of the more challenging, yet critical, learning issues I've noted is the difficulty therapists encounter in determining how to use ACT core concepts and other processes, on the fly, in a session, with a particular client across time. It is not unusual for therapists making this effort to innocently (and excessively) focus their attention on the content or form of behavior. Process and function are often minimally addressed or recognized, sometimes wholly lost. Although not necessarily a problem (we need to understand form as well), when this occurs, the gap between conceptual understanding and real-world work continues, and the deep experiential and heartfelt work is left behind.

This is not entirely surprising. We have had years of experience and extended learning histories all about this kind of focus—the content is most important; how the client feels and thinks is critical. We have extensive systems of verbal/psychological training built on this understanding. The social rules, or our verbal conditioning, are relatively evident: how you feel on the inside is what matters, and it needs to be fixed. Using the ACT model turns these rules on their heads. The function of behavior, not necessarily the content, is the relevant place of intervention; how you behave is what matters, is what's linked to values (for a more in-depth look at function of behavior, see Ramnero & Törneke's [2008] *The ABCs of Human Behavior*). But knowing this doesn't undo the years of conditioning. Our behavior in therapy can become overly regulated by the content (i.e., talking about thoughts and feelings is the rule). Switching to recognizing the function of behavior may be a bit more challenging than is often known, and it is easy to fall back on rules when you are unsure. Indeed, our therapeutic training can even support this overregulation (e.g., a good listener pays attention to content, whereas in ACT a good listener "reads" and hears more than just the words—they "listen" for function as well). We are experienced at it and trained in it, and it can come quite easily to us as therapists. This is in part why we can easily miss the function of behavior.

The content *and* function of a behavior are present in nearly every verbal interaction in therapy. From an ACT perspective, however, the problem occurs not only when the client's behavior is overly regulated by content, but also when our behavior as therapists is too regulated by content. Let's examine a simple social rule linked to good listening (a therapeutic must): be polite. This rule can be seen in a typical everyday interaction. If I ask in passing, "How are you?" and you say, "Fine," then you and I may have just had a "content to content" interaction. We continue passing each other without much else said; whether you are fine may not be relevant. Perhaps you have responded to the content of my question as a "be polite" rule (say fine when asked how you are doing), not its function (social connection). And my asking this question could simply be about following the same rule. Unfortunately, in therapy, we also can get caught in following this kind of verbal rule—be polite, or, in this context, listen well (e.g., don't interrupt, reflect what was said). And, in following the rule, we miss the story beneath the verbal behavior. We sit quietly and then reflect what was said. This is well and good and should be done as part of our therapeutic work. However, in sticking to this rule, we can "pass over" the client without much else getting done concerning addressing function. We might also miss the other processes (e.g., the body language and its meaning, verbal tone and pace, the person's experience across time, or this time compared to last time or other times), including personal feelings and reactions that inform function.

Sometimes getting at process and function in session means being willing to break some social rules (e.g., listen or be polite), "looking beneath" the content of verbal behavior. The example of asking how a client is, for instance, could function in a couple of ways. Let's assume momentarily that it functions to keep you and the client connected socially and interpersonally. You acknowledge the client and the client acknowledges you, and both of you pause to connect further with each other. This still might be content heavy. However, it can serve the broader function of developing the relationship.

Nonetheless, if ACT calls for me to get at function more routinely in session, then I will need to "look beyond" the words. So, with this example, I might look to see if body language or tone matches the expression "fine." I might look for eye contact and facial expressions. I might internally assess whether this "fine" is different from other "fines" the client has stated in the past. I might check my own experience in the moment to see how I feel, especially if the client's tone and body language don't match. If I don't "see" fine, then I am likely to break the social rule (and not simply accept the answer "fine")—I am likely to ask about what I see and hear in the client's tone or in the client's whole experience, rather than simply responding to the words. I will be moving to work "beyond the content" the client has brought to session and commenting on the change I notice between this time and the last (e.g.,

"You said 'fine' a little differently than you usually do"), or the experience between us (e.g., "You seem far away"), or my experience of the client's body language (e.g., "I notice that you are looking down"). And I will do this not just for questions related to how the client is doing, but also with respect to understanding behavioral patterns of the client throughout sessions and across time. In so doing, I'm not only letting the content of the session inform and guide my work, but also allowing the overarching, interpersonal, and intrapersonal process to inform my intervention, paying attention to how my and the client's interactions consequate and shape behavior. Is our process changing the function of verbal rules depending on context?

> ## Carlton's Reflection
>
> *This is why function can be tricky. Ultimately, if the function of behavior is determined by its consequences, then I think it is possible that the function of the client saying "fine" was to elicit more observational behavior from the therapist. In this context (therapy), when the client says "fine," the usual language game does not prevail. Maybe a better way of saying this is that, as an ACT therapist, I need to be able to choose to observe the functions of the client's behavior (including intra- and interpersonal responses) where this is useful, and to choose to respond according to a specific function selected from a range of functions, some of which may be established by content (or rules), and some by interpersonal processes.*
>
> Robyn: *Indeed, a client saying "fine" might well function to elicit more observational behaviors (more question asking) from me as the therapist. Understanding the context of the content will be necessary. How does the client say the word "fine"? Is it said sarcastically? Or does the client cast their eyes down when they say it? Or are they smiling and looking me in the eye? What do I notice when they say it? Or feel? The context of the event matters. Is the client following a rule—be polite—that functions to keep us socially connected, or is she "fine" so that we don't speak about emotions, its function experiential avoidance? My response will depend on my understanding of the client's behavior.*
>
> Carlton: *So, there is another way of viewing such exchanges, namely examples of a function-to-function interaction? If you are asking me how I am, and I am replying "fine," even though I may not actually be fine, I am likely responding to the "interpersonal functions" of your question. We are playing a language game: I ask you how you are, you say "fine"; that's the protocol. This is about you communicating your care for me, and me acknowledging that. If you asked me in passing, "How are you?" and I replied, "The kids are driving me crazy, I haven't slept properly for years,*

> *people at work are stressing me out,"* then the function is different because it elicited a different response. But this is still a function-to-function interaction. It's just this time the function was influenced more by the literal content of what was said rather than established social protocols.
>
> Robyn: Agreed. The act in context is incredibly important.

When referring to content, as many who are learning ACT will know, I am referring to the actual thoughts, feelings, and sensations an individual is experiencing. So, for example, if I am looking at the content of a client's thoughts in session, it might be a story she is telling me about her past that involves the memories of an event and the thoughts that she might be saying to herself regarding or related to those memories. She might say, "I was ultra-responsible as a child. My mom was always sick, so I had to take care of my brothers and sisters. My value was in what I did to help out. I felt guilty if I wasn't helping my mom. I feel guilty now if I am not constantly doing things for others. I'm exhausted and feel like a failure when I can't meet other people's needs." First, let me say that the *content* of this story is important. As a clinician, I want to know this information. I am quite likely to ask questions about and get more examples of these kinds of memories and understandings of her experience. I am interested in the story—its form— of what she has done in the past. The target of my intervention, however, is going to be about *function*; I will be targeting a pattern of unhealthy behaviors—a class of behaviors that function to avoid guilt.

> ## Manuela's Reflection
>
> *To speak to the content and function in every conversation, let's look at the example of the woman who experiences guilt more closely. It is here that we can see where therapists might struggle. The behavior of the listener can be more regulated by the content of what he or she is listening to rather than by its function. That is, the therapist might begin to ask what feeling like a failure is like and spend time asking about and working on how difficult it is to meet everybody's needs rather than focusing on avoidance of guilt.*

Because we are socialized in our clinical training mainly to respond to content, it is easy to get caught in it. Really easy. Indeed, it is so easy that some therapists are not even clear about whether it is happening. How might you know if you have been caught by the content? A few of the "tells" of being caught in content are relatively easy to identify: (1) only responding to what the client *says* instead of also responding to other aspects of the client's behavior (e.g., body language, tone of voice), or to interpersonal process (such as noticing something that is happening between the two of you), or to your intrapersonal process (such as noting an emotion the client is evoking in the moment), or to other functions (such as avoidance, an overarching and ongoing process or pattern of behavior linked to case conceptualization); (2) overexplaining ACT and its concepts rather than working on experiential processes (the six core processes); and (3) getting lost in trying to figure out what is the next best ACT technique to use (stuck in rule following).

Other ways of getting caught are subtler. For instance, getting caught in content might look like a never-ending assessment. The therapist asks questions throughout session that call for content-oriented answers. It might seem like the therapist is heading somewhere, but no intervening ever seems to come from the questioning: How did you feel? What did you think about that? Then what happened? How did you like it? That must have been hard; how did you handle it? Each is a legitimate question. But if the therapist continues to ask some form of these questions throughout the session, with no link to a process or the purpose of behavior, then it's likely function will never truly get addressed. This issue might be trickier when the question appears to be ACT consistent, such as "And how does that work for you?" This question can be overused and inadvertently become nothing more than a content trap. The tell here is the frequency and placement of the question. It can be asked too many times and at places where it doesn't fit, or it can function to inadvertently dismiss the client or evoke shame.

An even more challenging "miss" related to being caught in content rather than function occurs when the session is loaded with other ACT-like qualities, but the content is still the only level of interaction in the room. For instance, the therapist might be diligently asking the client to notice what they feel after each moment of telling a story or sharing a memory, but the purpose of noticing is never contextualized; noticing is done for its own sake.

Paying attention to these subtler therapist actions, and staying connected to process and function, is part of the process of becoming skilled in ACT. The work done in ACT involves noticing the story and how, when it shows up, the client feels compelled to act on it (e.g., being impulsive, not getting out of bed). This noticing, this slowing down, provides the opportunity for change: being open and aware of your own plus the client's experience, setting the stage for choice. Willingness is then available, right there in the room, thus changing the client's relationship to their content, "freeing" them to live; and perhaps changing the therapist's relationship to their content, freeing them to break the rules—responding to function rather than form, engaging in process rather than content.

> ### Carlton's Reflection
>
> *I have found that noticing how entire stories can compel a client to act can often be very useful in therapy. Doing so provides freedom for the therapist and client to pay attention to the macro level of broad narratives, as well as the micro level of individual thoughts.*

Content Matters: It's Okay to Listen Well

I want to comment a bit more on the client's stories, the content of their life, in general. I am routinely asked whether it is "okay" to listen to a client's story (content) as an ACT therapist, even after noting that paying attention to content is important as well. The answer is an enthusiastic "yes." There are fair amounts of time in a session that can and probably should be given over to the client's story about their life and struggles. It helps the clinician to understand the client's learning history and developmental milestones. It can help the clinician to know the particularly difficult aspects of the client's upbringing. It can help with case formulation and predicting what the client might do or say. The therapist's behavior of listening to a client's story can have a valuable function in the therapy room. But most importantly, it can serve to help the client feel heard, understood, acknowledged and "seen." So again, yes, please do deeply listen to your client's stories; just don't stop there—don't get caught in that place and sacrifice function to form.

> ### Manuela's Reflection
>
> *Recognizing that it is okay in ACT to listen to clients' stories was such a relief! Sometimes when we get a new rule (e.g., pay attention to function), the pendulum can swing too far toward the rule. If it first swings too far toward listening to content, it can then swing too far toward finding the function. In my own developmental process with ACT, I became a "function fundamentalist." When I look back at my therapy sessions during that time, I realize that I was less willing to listen to the stories of clients—rushing instead to find the function of the behavior, getting impatient when clients "talked too much." This made my encounters with clients less human. Emphasizing that we can listen to clients' stories is really important. It is not one versus the other, function versus content. It is a more flexible balance between the two that supports both a pragmatic behavioral intervention linked to function and a meaningful therapeutic relationship.*

Growing in Your ACT Practice

As you begin to grow in your understanding of targeting function of behavior as the point of intervention, I still invite you to entertain *no arrival*. Even for the person who has done a fair amount of training and says, "Yes, I get it, this makes sense," the translation from getting it to doing it may remain a challenge. Many learning ACT and trying to master its processes will go out into the therapy world and do session after session of content-based intervention, responding over and again to societal rules (e.g., be polite), psychology training rules (e.g., have compassion no matter what, leaving wisdom behind), or "ACT rules" (e.g., do this exercise in session four; use defusion whenever you see fusion; use present-moment work whenever you think the client might be having an emotion; use the clever intervention you saw demonstrated in an ACT workshop).

Manuela's Reflection

How would you advise the therapist on learning how to notice when they are focusing on content or following ACT "rules"? Is there a way to help them be aware and shift to something deeper in their work?

Robyn: *Yes, and the answer takes us back to the importance of experiential learning. Working on personal openness, awareness, and engagement will be an essential ingredient in noticing when the therapy has become content focused. Other ways of noticing might include reflecting on how much you are talking in session, whether you are stuck and the story is remaining the same over time, and whether you find your sessions a bit lifeless.*

I recently had a supervisee state that while the client was talking, he was running through ACT techniques in his head, searching for the next thing to do. However, this activity wasn't organized around what the client's behavior was showing; it was organized around the content of the metaphors and exercises he had been studying. He was talking to himself as follows: *Should I do a defusion exercise, maybe milk, milk, milk? No, wait, what is the client saying? Oh, she is talking about an incident with a family member. Should I do a values exercise? Okay, focus. Maybe I should have her show up to what is going on inside her skin? Do some present-moment work, or maybe help her see family member as a role and explore self-as-context?* And around and around it went. The therapist was sitting silently through a large part of the session, but it wasn't silent in his head; he was desperately searching for which technique he could do…because he really wanted to say he was practicing ACT. In fact, he had been to multiple trainings and could talk to you about ACT techniques and what they are designed to do quite well. He had numerous ACT books on his shelves and had read through many. He had been doing therapy for years. The trouble wasn't knowing the material. The problem was that he was checked out of the ongoing process across time, he was checked out of the relationship process, and he wasn't even present to his own intrapersonal process. He had lost the thread of the client's struggle inside of the broader context of therapy and had gotten caught up in the thread of figuring out ACT.

Addressing this gap, from getting it to doing it, is part of the journey of mastering and participating in the heart of ACT. Opportunities to implement ACT require slowing down and showing up. In the example above, I worked with this supervisee to slow down and listen to the client. From that space, he was asked to notice his internal reactions to the client and his desire to help. I then asked, "If you were to hold this client as whole and acceptable, and in such a way that no matter what content arose, no matter what painful thoughts or feelings showed up, you were to communicate to her simply that—simply acceptance—what might you have done?" Here the supervisee noted that he would probably sit quietly and listen, but more importantly, he noted that his head would probably not be running all over the place searching for the perfect ACT technique. This was a good start. Slowing down, for even a few seconds at a time, becoming aware of overarching, interpersonal, and intrapersonal processes, will assist. This involves, in part, paying attention to when, where, and what you use, as well as how you combine it with other exercises, while simultaneously interweaving it into the client's presenting issue in such a way that it fits what they have shared with you in the past and moves them forward in the session and more broadly. Remember it is possible. Moving from being trained in ACT to implementing ACT in the real-world clinical setting and from an experiential place and heartfelt connection is personal and process oriented.

Despite an understanding as well as an emotional connection to ACT, or even a real sense that it fits for you as a therapist, the leap across the gap from training to fully engaged mastery takes time. It requires connecting to your ACT therapeutic stance, a topic we explore in chapter 10. Filling the gap (i.e., knowing and attending to the distinctions between process and content and between function versus form) requires at least two things: confronting your tendency to respond to content and utilizing contextualization as a guide.

CONFRONTING YOUR RESPONSE STYLE

As noted, we are highly trained to respond to content. It is what we are often doing when we are talking with others. This well-learned response style will need to be confronted. Self-awareness regarding your own motivations and stories about your behavior will be important to contact and then hold lightly. An essential part of pushing forward in the development of your ACT skill will be knowing when and where you are "buying" your content, including rules related to ACT, and knowing when you are attaching to certain concepts about yourself, fusing with your own mind, and falling into "choiceless" action. Recognizing this in yourself will assist you in recognizing it with others. Ask yourself, *Where do I reach my limits of what's acceptable concerning my experience, and what do I do when I get there?* If you answer, *There is no limit to what's internally acceptable*, take a moment to reflect on your most significant struggles, your most painful moments, and see if there needs to be an honest acknowledgment. Look carefully and note if there might be things that you would like to go away. Whether it is fear or jealousy, whether it is anger or desire, we all bump up against places we wish we didn't have. Are you able, in these darkest moments, to be with yourself openly and compassionately? Are your values clarified, both broadly about your personal life and regarding how you want to be as a therapist and in the therapy room with your clients? Do you indeed recognize that there is no arrival? That acceptance as an outcome and ACT as an intervention are not the ultimate answers? Accepting and committing to a valuing process in life is just that: a process.

This kind of self-knowledge, this level of personal awareness, will be a guide. It will not only help you as you step forward in this weird, wild, tragic, lovely thing called life, it will also help you help your clients. You will be able to empathize with them at another level, a level that is about the never-ending struggle that human beings encounter. Life is a rising and a falling, a coming and a going. If you are free from pain now, just wait a bit. Genuinely, authentically, recognizing life as process and your own impermanence can translate into thoughtful and fuller work with your clients. It might lead you, as well as guide you to direct your clients, into full engagement in this very temporary thing called life. Recognition here supports the urgency needed in working with clients, but not a rushed urgency, rather a passionate urgency. An urgency, to the best of one's ability, to not let another moment pass when the struggle-filled, painful, historical story dictates how you live.

UTILIZING CONTEXTUALIZATION AS A GUIDE

Knowing process and function with your mind and getting a "feel" for it as it unfolds in a rapidly moving therapy session are two different things. Not only can therapy sessions move very quickly, but the process(es) can also move quickly. The tears and emotion of a client can rise and fall in mere seconds. The same behavior can have multiple functions, and different behavior can serve the same function. This means what should be done can feel like a moving target or an unknown. Indeed, the question "How do I know how and when to respond to process or function?" is best answered with "It depends."

This leads us back to contextualism, which is helpful in guiding these kinds of clinical choices. Knowing what process to focus on or where to intervene on the function of a particular behavior depends on the context and history of your interaction with this client in a specific session, and even a moment in session, and the function of a particular behavior. Being aware of this kind of minutia while trying to implement the six core processes may seem nearly impossible, but the good news is that it is imperfect and doable. There are many ways to respond that can make a difference for the client.

Becoming fluid in ACT may also mean taking a risk with others. Getting supervision or consultation, working with a group of therapists, presenting cases, and tuning into your own process can make a difference. If you are bold enough to have supervisors or consultants listen to or watch recordings of your sessions, despite increasing your anxiety, it can shorten the gap between getting it and doing it—with heart and wisdom.

REFLECTIVE Practice 6.3

First, review and explore the questions from the above text:

- Am I able, in these darkest moments, to be with myself openly and compassionately?

- Are my values clarified, both broadly about my personal life, and regarding how I want to be as a therapist and in the therapy room with my clients?

- Do I indeed recognize that there is no arrival? That acceptance as an outcome and ACT as an intervention are not the ultimate answers?

Second, ask yourself, *Where do I reach my limits of what's acceptable concerning my experience and what do I do when I get there?* And, if you answer, *There is no limit to what's internally acceptable,* take a moment to reflect on your most significant struggles, your most painful moments, and see if you need to make an honest acknowledgment. Look carefully and note if there might be things that you would like to go away. Then ask yourself again:

- Am I able, in these darkest moments, to be with myself openly and compassionately?

- Are my values clarified, both broadly about my personal life and more narrowly regarding how I want to be as a therapist and in the therapy room with my clients?

- Do I indeed recognize that there is no arrival? That acceptance as an outcome and ACT as an intervention are not the ultimate answers?

The Heart of Now and the Door to Next

The heart of ACT lies not only in growing your mastery of a complex set of skills, but also in cultivating an awareness of multiple processes that are coinciding, linking them together in a shared understanding guided by purposeful activity. Observing, tracking, and attending to the overarching process of therapy across time, the interpersonal process, and the intrapersonal process are all part of what allows the therapist to contact, understand, and intervene at a functional level. Knowing your own patterns of behavior as well as in-the-moment responses, the client's patterns of behavior as well as in-the-moment reactions, and the interplay between the two supports the capacity to have a responsive and flexible ACT practice. I look forward to exploring these more fully in the coming chapters as we take a closer look at how these processes unfold in therapy. Let's turn now to chapter 7, where the processes defined here are more fully explored.

CHAPTER 7

Overarching, Interpersonal, and Intrapersonal Process in Practice

Change will not come if we wait for some other person or some other time. We are the ones we've been waiting for. We are the change that we seek.

—Barack Obama

One of the more powerful treatment aspects of ACT is its focus on process and the function of behavior. These are intimately intertwined. Extending beyond the six core processes of ACT and beginning to explore the different levels of process as well as the therapeutic stance is an integral part of moving into the heart of ACT. Translating technique into fluid implementation involves, in part, being aware of and connecting to the contextual, process-oriented layers of therapy. If we accept that the work done inside of an ACT session is about conceptualizing and influencing the ongoing purpose of behavior in the context of the client, the therapist, and the relationship between the two across time, then the work of implementing ACT well will involve embedding the six core processes into these larger process layers, as briefly defined in chapter 6. Bringing these layers into focus by exploring them in practice is where we will spend the bulk of time in this chapter.

Science, although absolutely relevant, does not fully capture the wisdom and felt experience of being human. The heart of the work reaches beyond science and into the ever-changing flow and connection between you and client. To see into the heart of your client with understanding, kindness, and care by recognizing what flows within you as well as between you and your client, and to see how this flow changes across time, creates a therapeutic experience where two separate people merge into humanity, establishing us not as "things" to be manipulated but as beings, experiencers to be encountered.

> ## Manuela's Reflection
>
> *Why is it that process-oriented psychotherapy isn't equivalent to working with the six core processes of ACT?*
>
> Robyn: *Working with the six core processes of ACT is part of a process-oriented therapy, and indeed, they are processes themselves (i.e., ongoing, continuous) with procedures that play a significant role in therapy. However, these processes do not occur in a vacuum. They are intimately interwoven with other ongoing and continuous processes that literally shape our behavior and define our therapeutic interactions and outcomes.*

To illustrate the topics in this chapter, we will explore how overarching, interpersonal, and intrapersonal processes play out with a particular client and how to take these processes into account when conceptualizing the case.

Case Conceptualization: A Clinical Example

Darrell is a thirty-seven-year-old businessman who presented to psychotherapy with health anxiety. He had been terrified following a routine medical doctor's visit wherein he learned that he had cancer. The cancer had grown to such a degree that he was whisked into surgery just two days after the routine visit. Following the surgery, he began to imagine that cancer was growing in other parts of his body, and he was often showing up at the doctor's office asking to have more tests. These experiences reminded him of a time when he was hospitalized as a child with pneumonia, another time when he was quite ill and almost died. He finally decided to see a therapist to assist with his anxiety. At the first session, Darrell told the therapist that he worried about his health on a regular basis. Darrell indicated that he couldn't sit still for hardly any reason. If the therapist asked him to focus on the present moment by paying attention to a part of his body, he would say, "I can't, I just can't," and would refuse to spend time reflecting on his bodily experiences.

A relatively good case conceptualization might include noting that the client is trying to control his anxiety as well as any other painful emotional content that might show up around getting a diagnosis of cancer, being ill, other health issues, or death. Being aware of his "I can't" statements might be couched in terms of fusion. A closer look would include noticing that he is also seemingly avoiding things that have to do with his body in the present moment as he may wonder if any of the tiniest bodily sensations are a sign of cancer or other illness. Exposure using willingness and acceptance exercises in combination with defusion techniques might well assist the client in decreasing his struggle. This brief assessment makes sense in the initial conceptualization of the client.

Another kind of curiosity around case conceptualization will be helpful in extending the ACT work beyond the above understanding what is happening for the client. Pushing forward into something more fluid and responsive will require additional work. Rather than simply looking at what is happening for the client in terms of the six core processes in ACT (e.g., willingness and defusion in this case conceptualization example) and their associated techniques, the therapist will also want to focus on ACT processes in terms of overarching process across time, interpersonal process, and intrapersonal process.

Overarching and Ongoing Process

Conceptualizing the *overarching and ongoing process across time* allows the therapist to entertain a broader description and understanding of the client's patterns of behavior, providing more systematic and coherent insight that can heuristically guide inquiry as the sessions progress. This approach makes therapeutic information tied to ACT pathological processes (e.g., fusion, conceptualized self, experiential avoidance) more accessible. Intervention under this circumstance is more fluid; when guided by a systematic and coherent understanding, the therapist can more readily act, incorporating intervention more responsively and methodically. In Darrell's case, this means using willingness and defusion steadily as processes with the goal of helping the client to "see" "mind" and thinking, not just expressing the thought *I can't*. This kind of process looks different from searching for a technique. Rather, it is based on a larger "compass" that guides the work and is responsive to change across time. For instance, rather than simply doing a repetition exercise with "I can't" or saying, "Thank your mind for that," when Darrell says, "I can't," the therapist can reflect on mind more broadly:

- Noticing the role of "I can't" in the client's life across time (bringing in self-as-context work and perspective taking).

- Connecting to and becoming aware of the pull to encourage and refute "I can't," one mind to another (therapist's mind to client's mind), with "Yes, you can."

- Pointing to the emotional experience that comes along with "I can't" in the room and across history.

- Noticing how "I can't" will play out in the future, predicting its enslaving qualities and its capacity to pull the client away from values.

- Speaking to the "wall" between client and therapist that "I can't" seems to erect (interpersonal process) and the therapist's emotional experience when "I can't" surfaces over and again in therapy (intrapersonal process).

- Looking for the other ways "I can't" shows up in therapy and in life (e.g., not completing homework).

In considering each of these possibilities, the clinician's "flexibility" compass can be pointed in the direction of freedom from mind, not simply defusion from words.

Furthermore, regarding the ongoing process across time for this client, multiple considerations emerge linked to other core processes. Not only is Darrell currently working to avoid his fear of cancer, but he is also caught in a story that stretches across time and is related to his conceptualized past and future selves as well as his experiences of change in the past and up to now. His present moment is lost. Darrell was frightened, wondering about and facing his own death in a very short period of time—at one point in time he is "healthy" (before the diagnosis), and at another point in time he is suddenly not (after the diagnosis). Self-conceptualizations from one point to the next were rattled, conceptualizations of his future undone.

Life, from moment to moment, is unpredictable. The therapist can explore patterns of unpredictability, pointing to this as an experience that he had in the past, and as a possible experience of the future. Working on flexibility from this perspective makes sense. Responding and adapting to an unknown future are part of enhanced flexibility. Holding any conceptualization of the self lightly will be helpful across time, noticing its alluring power or even positive qualities. Whenever the therapist detects past, present, and future senses of the self as "truth," he can turn to the pattern of unpredictability once again as a way to gently free the client from a "known" identity. In exploring unpredictability, the therapist might invite the client to be willing to feel what he feels under such conditions.

There is another piece to this client's story that may prove a valuable heuristic informing the overarching and ongoing process. A part of the client's experience when looking forward into the future is not unpredictable, but entirely predictable indeed—his own death—not its time, but its inevitable occurrence. If we harken back to chapter 4, you will recall that awareness of death can move us from a state of wondering "why" we live to a process of engaging "how" we live. It can catalyze change and is an ally of meaning. However, touching on this experience through a diagnosis of cancer is likely to be anxiety provoking. Allowing that anxiety to steal his current life is yet another thing. Fusing with "I can't" in the moment is a robbery of life in action. Ongoing and varied questions about "meeting" life, in the context of death, may potentially be powerful explorations for this client. Values-based and committed action activities can be weaved into these client–therapist conversations in a way that touches the past, touches what is happening now, and touches how the client wants to engage the future—cancer or not.

Working in this fashion, sharing with the client insights about how "I can't" has functioned, there and then, and here and now, allows the client to expand his sense of himself, fluidly incorporating self-as-context and perspective taking without needing to use a specific technique, but instead engaging in layers of process. Finally, each interpersonal transaction contains something of the larger whole. Therapist and client interactions will contribute to ongoing process and change across time.

> ## Manuela's Reflection
>
> *In reading the passage on overarching and ongoing process, it sounds like there are multiple hierarchical processes (see Villatte et al., 2015, for more information on hierarchical framing). So, you can tackle processes at one level as they relate to the hexaflex. As well, it sounds like you can work with more hierarchical processes that include the other three processes you mention. There is a hierarchy of processes that we can look at. What is the benefit of looking for the overarching process?*
>
> Robyn: *When working from the ACT perspective, I don't simply want to focus on process in the session; I also want to keep in mind the interplay between the client's stated values and the arc of the therapeutic work. This way I can stay oriented to what is happening now as well as to where we are going next. It contextualizes the work I am doing and is part of a safeguard against doing exercises without purpose.*

Interpersonal Process

The example of Darrell's case speaks to issues of unpredictability and death that may serve as organizational understandings of the work in therapy (but are held lightly in case a shift is needed), all brought into the primary place wherein life experience is construed—*the interpersonal relationship*. The interpersonal transaction should be highly emphasized as part of the ACT growth in mastery and heartfelt process. It is the stage where problematic life experience is played out and responded to. Therapist and client are engaged in a joint process of discovery and movement linked to change over time, but it is playing out in the current interaction. Hopefully, the therapist and client are collaboratively engaging in the process of experiencing, whether it be contacting moments of joy or moments of pain, in the service of positive or values-based change.

Here the therapist can respond by bringing compassion to bear in moments where needed, and other consequences as called for. Working from the interpersonal process perspective is an "on-your-toes" deal. Recognizing process and function requires extraordinary awareness of what is happening between you and the client. The client will be inadvertently shaping you, just as you are shaping them.

Continuing from above, a proper ACT case conceptualization might include noting that Darrell is avoiding anxiety when the therapist invites him, in session, to sit with bodily experiences. It might also include being aware of the back-and-forth content exchange and looking to see if that exchange serves to continue the avoidance. If so, this should be noted, and both client and therapist should work on a defusion process. The therapist would model acceptance and speak to the client from a genuine, compassionate, and sharing point of view, perhaps self-disclosing about his own anxiety, at appropriate moments. Modeling willingness as well as focusing on present moment, where both sit in awareness of experience, would be consistent with the ACT model.

Becoming more fluid and responsive will require further exploration of the interpersonal process. Mutual perceptions of and reactions to what is being said and experienced are a part of this fluid and responsive style. This might include questioning the client about his reactions to the therapist's comments and emotions. It might also involve noticing the opportunities to draw parallels between the client's interpersonal difficulties and any difficulties that may be occurring in the therapeutic relationship. The interpersonal process focuses on the features of client and therapist that are unfolding in the story, speech quality, body language, and how the therapist is interpreting these and responding to them in a reciprocal, dynamic process. This may require the therapist to hold not only what is occurring in the moment, but also what has happened in the past between client and therapist as well as what could potentially happen in the future.

Specific to Darrell's case, the therapist may want to attend to his resistance (i.e., fusion with "I can't") and the other verbal and nonverbal responses he displays in any moment that indicate resistance. As well, the therapist may want to reflect on these experiences and how they impact the therapist himself while asking how this kind of impact influences Darrell. For instance, imagine that the therapist asks Darrell to be aware of any anxiety he has in his body. And just as the therapist asks this question, Darrell has a little flinch in his left shoulder and then states strongly, "I can't; I just can't." The therapist might then respond from an interpersonal and ACT-coherent perspective, stating something like "I noticed that when I asked you that, there was a little flinch in your shoulder, and I suddenly felt like I shouldn't ask you to do things like notice your body. I am wondering, first, if you noticed the flinch, and second, if that seems to be telling me to back off." This response incorporates noticing with interpersonal impact. This is richer than simply asking the client to defuse (e.g.,

"Thank your mind for that"). The therapist might then go on to ask Darrell what it is that he experiences when he (the therapist) is noticing things about him beyond his words (e.g., the therapist is noticing Darrell's body movement). This intervention intertwines defusion (e.g., noticing, physicalizing—"telling me to back off") with perspective taking (e.g., "What is it like for you to know that it is telling me…?") and is linked to the interpersonal quality in which the experience is embedded. The therapist is not simply focused on the story Darrell is telling (i.e., I can't); he is also focused on body language, perspective, and perhaps his own internal response. He also appears to be working on personal willingness as he is being vulnerable; in other words, the risk is that Darrell might get upset about the therapist noticing his body and not responding to the words "I can't."

From here, the dialogue could go in many different directions. Darrell might say, "That *is* me telling you to back off (pointing to shoulder)" or "What flinch?" or "It doesn't matter what it's like for me" or "Really, I've never noticed that before." Or he might get quiet and sink deeper into the chair, or he might even repeat himself, not having heard the question: "I can't, I just can't." The therapist's response, *your response*, will matter, and recognizing the flexibility within the interpersonal system, rather than getting locked into a rule, will shift your therapeutic experience, as well as that of the client, leading to something that feels more open between the two of you…or something that feels more closed. Either is there to be attended to.

Intrapersonal Process

Finally, ongoing case conceptualization or functional analysis will involve being aware of your *intrapersonal process*. Noticing what you are experiencing in relation to what the client says and does informs what happens in the therapy. There are two potential challenges. One is linked to the capacity to experience, notice, and speak to your internal experience and body language. The second is tied to knowing *when* to talk about your intrapersonal experience and when not.

Noticing your internal experience during session means attending to what you are feeling, thinking, and sensing in the context of your clinical work with a particular client. Some therapists are quite good at this but never use it; others have more difficulty and feel unable to say what they are experiencing internally in response to a client. Still others use this kind of information in a well-timed and placed fashion. The goal is the last.

On one end of the spectrum, the therapist's internal experience might genuinely match that of the client, and the therapist may have strong feelings of empathy that they consider and perhaps share. On the other end, the therapist might have no reaction, might feel nothing, which they also might share, and which may even be valuable to the client. Each piece informs the overall case. It is important to ask, *What does this client engender in me?* The therapist can then explore whether it is useful to bring the answer to this question into the therapy session or not. Overall, it is important to remember that emotions are intimately involved in experiences such as closeness and trust, both of which are part of successful relationships (Greenberg, 2015), including the therapeutic relationship.

As well, Hendricks (2009) suggests that in-depth experiencing is predictive of favorable outcome in therapy. The client cannot merely talk about what is happening in a detached manner; it appears that they also need to encounter emotion richly or deeply, viscerally experiencing and acknowledging it. This is not to say that emotional contact alone is adequate to promote change; the interplay of emotion, thought, and action is complex. However, therapists who are unwilling to reflect on, be aware of, and engage their own emotional experience may have difficulty inviting clients to engage theirs. The therapist plays a crucial role in the process of opening to experiencing in session by holding a stance of acceptance toward their own.

Indeed, emotions play a number of critical roles in our lives (Greenberg, 2015). For instance, emotions are a signal to ourselves, allowing us to be aware of how things are going in the environment and in relationships. They may assist in organizing our behavior regarding values-based action, playing a pivotal role in pointing to what matters, whether the emotion is pain or joy. In any case, our emotional reactions are part of the therapy session too. We are automatically responding to sights, sounds, smells, another person's intention, or the words being stated by the client. Attending to one's own body, being aware of the flow of sensation and emotion, will be part of the therapeutic experience. The utility of being aware of emotional experience then will be in recognizing when to act on these experiences and when not. This choice will depend on the case conceptualization and what will be useful to the client in making a change in their own life.

The therapist might consider self-disclosure in the service of the client, for instance. This might be done to join with the client (i.e., linked to relationship building), or it might be about empathy and connecting to the client from a felt place. A therapist might also consider self-disclosure when they want to provide information to the client about their impact on the therapist, linking this to broader patterns of behavior that are targeted for change.

Imagine that, in Darrell's case, the therapist is trying different intervention strategies week after week to set the context for exposure to bodily sensation. And week after week the story is "I can't." The therapist is beginning to feel blocked in forward movement and is noticing that he is wondering about Darrell's willingness to make a change. He is feeling frustrated with the lack of progress and the stubborn way in which this story is hanging on. It might be important to carefully share this experience with the client while being aware of how this might parallel what Darrell is experiencing. Darrell, too, may be feeling frustrated with himself and the progression of therapy. Using the experience of frustration in a nonthreatening and authentic fashion could potentially pull both out of the stuck place they find themselves in. The therapist might say, "I have been noticing, since we started working more deliberately on anxiety, our progress has slowed. I am experiencing feelings of heaviness in my chest, a kind of blocked frustration. Are you noticing anything like that?" (By the way, several of the ACT processes are present in this experience and are modeled by the therapist: committed action, willingness, defusion, and present moment.)

Darrell could reply with any number of different responses. Whatever way he replies, the therapist should again attend to what he feels and decide whether to share this next experience or not, asking himself, *Will this benefit the client or just me?* If the former, the therapist may end up talking about the sadness of being stuck, the two reflecting together (interpersonal process) on the pain of a life seemingly held up by anxiety. Both can focus on acceptance of this sadness and agree to continue to press forward with the difficult process of exposure, opening to feeling. If the latter part of that question is answered with a yes, then it makes sense to sit quietly and listen. But the therapist shouldn't stop there; he should explore his reaction by reflecting on it individually or with a colleague to determine if it does have relevance to the case and, should it be reencountered, needs to be examined in session.

Ultimately, the intrapersonal process involves both head and heart. Being aware of the ongoing flow of your internal emotion, thought, and sensation, and determining when and how to use it in session, requires awareness to the embodied experience as well as risk taking concerning the same during session. Engaging in a personal practice that cultivates this awareness is fully part of ACT's heart and growth in mastery. I don't want to say that therapists must do this, but I do take a firm stance: knowing your emotional experience and being able to tell when you are triggered into an emotional reaction that is not likely to be helpful to the client versus a response that will assist the client in recognizing their impact on others and in moving forward—this is essential. And this can only come with awareness to the intrapersonal process and a willingness to put it in session.

Focus on Function: A Clinical Example

Dan is a forty-year-old male who presented to therapy with feelings of extreme loneliness. He reported that he was performing poorly at work, that he wasn't "eating right," and that he had been desperately wanting a partner for more than ten years. He had been previously married and had two children, neither of whom he was close to or made much effort to remain in contact with. He reported extended periods of depression, primarily based on feeling lonely. Dan spent many hours at home reflecting on his situation, searching through his past, and wondering how he was going to overcome his loneliness. He presented to therapy wanting to understand why he felt the way he felt, wanting to get insight into his problems so that he could be free from them. He was stubbornly persistent in his attempt to gain understanding. This understanding, he surmised, would lead him to a place where he felt more competent and thus able to get a partner—ultimately bringing about the resolution of his feelings of loneliness. In this effort, he was doing meditation for one to three hours per day. He was also attending weekend silent retreats and longer retreats to reach the enlightenment he so desperately craved and thus to achieve partnership.

Overarching and Ongoing Process

Overarching process across time as it relates to function in this case may be relatively easy to spot. Dan is trying to avoid feeling lonely but has been clearly stuck in an "if you don't want it you have got it" position. His avoidance strategies, sitting at home reflecting and trying to get insight, are functioning to perpetuate the feeling of loneliness. He is fused with a concept related to gaining understanding as a means to feel more competent, freeing him to act. The "catch" is the mindfulness and meditation work.

In looking at the function of behavior across time, it's critical to consider Dan's longstanding effort to get reflection, insight, and meditation to "work" and how these strategies have been highly costly to Dan over time, given his desired outcome for therapy (getting a partner in the service of not feeling lonely). Ten years of struggle is the first tell. His "getting insight" behavior is highly entrenched, and it is likely that creative hopelessness (see Hayes et al., 2012) will need to be deeply explored. Creative hopelessness may even come to be considered an ongoing process itself, being used, engaged, and reengaged each time Dan turns to understanding and insight to overcome loneliness, especially if he continues to meditate while continuing to take no action (e.g., not getting out of the house, not contacting and building relationships with his children).

This scenario may be particularly challenging for the therapist in two ways. First, if the therapist is unwilling to experience the ongoing anxiety linked to a steady dose of creative hopelessness (i.e., undermining the control agenda), it may be difficult to disentangle the client from the unworkable agenda. Second, if the therapist, in the process of wanting to help the client move toward willingness and defusion, is hoping mindful awareness will be part of the journey, another difficulty may ensue. The client's practice of meditation, in and of itself, may not be problematic. However, using it in the service of insight to understand and overcome loneliness is another issue.

Indeed, Dan has lost contact with the purpose of mindful awareness and is using it to obtain a desired outcome.

Here is the trickier part. Imagine that Dan is beginning to "see" how his meditation practice isn't working via the creative hopelessness work and he starts to talk to the therapist about giving it up—letting go of the idea that meditation is going to help him overcome his loneliness. He notes that he is going to decrease the amount of time he spends meditating. He continues by saying that he now understands that meditation doesn't work this way; he has had the insight that meditation is about observing, not about obtaining a particular outcome. Unless he also reports that he is now getting out of the house and connecting with people regardless of feelings of loneliness, this new insight may function to keep him stuck. He may still be caught. Letting go of obtaining a particular outcome while meditating is now the "new" practice that is going to free him to be able to act. If the therapist is caught in content, this may seem like a good outcome; however, if the function is explored, the therapist will see that perhaps more creative hopelessness is needed and is part of the ongoing and overarching process in the arc of the therapy.

It is important to recognize that Dan is changing his behavior, but the new behavior has the same function—his new insight is letting go of attachment to meditation to obtain a particular outcome, so he will meditate with no desired result, *and* this will give him the insight that he needs to overcome loneliness. The therapist, engaged in an overarching process across time, will be looking at the function of behavior, including new behavior, to see if it is workable with respect to values. It is also important to note that new behavior, even if it still functions to sustain avoidance, is potentially a good sign. Dan's ability to let go of meditation as conceptualized and decrease its frequency should also be woven into the therapy as a strength, but it's his ability to shift his actions more broadly such that they are linked to his values, not the actual behavior he changed to (less meditation), that matters.

Interpersonal and Intrapersonal Process

Tracking the relationship during therapy sessions with Dan will entail a fair amount of awareness to *interpersonal process*. In particular, the therapist will want to be able to hear, see, or "feel" when Dan is experiencing loneliness during a session. If Dan is highly persistent in trying to gain understanding, opportunities for the experience of loneliness may be few. In this case, the therapist may want to turn to how understanding is working, right now in the room, in helping him to overcome loneliness. If Dan reports he is not experiencing loneliness in the room (as he is too busy in his head working on understanding), the therapist may use her *intrapersonal process* to point to what is happening. Dan spends so much time in session reflecting on his inability to get a partner and how he now understands that the meditation was wasting precious time, he doesn't get the chance to feel. In his working out how all of these are interconnected (and perhaps doing so in a persistent way), the therapist may get "lost." She may be left feeling alone and perhaps frustrated, in the room, during a session. Here, in the service of assisting the client to feel what he has been avoiding—loneliness potentially—the therapist may interrupt Dan and say, "I am feeling lonely right now." The interpersonal and intrapersonal are brought to bear simultaneously, with the function of disrupting the persistent "getting understanding" behavior. Process can also be brought into awareness. The therapist might say, "I have been sitting here listening for quite a while, and I have noticed that even though you are looking at me and talking, it is almost as if I am not here. Your mind has you, and I have disappeared. I noticed loneliness there." Loneliness is now in the room. A patient silence may be called for next. The opportunity to explore the function of ongoing verbal behavior now present, exposure to loneliness the feeling is now possible.

Again, many different things can happen following this kind of process-oriented interpersonal comment. The client may keep talking; he may look puzzled and try to understand (a likely outcome); he may reflect on what has been happening; he may even connect to a small bit of what he does to create loneliness. The therapist's next move will be based on what Dan does at the moment and how it functions according to what the therapist has understood about his behavior in the past and in the moment, and where it is they are trying to go: overarching and ongoing, interpersonal, and intrapersonal speaking to the function of behavior across time.

REFLECTIVE Practice 7.1

Audio- or video-record an ACT session. Listen to (or watch) the session. Explore the overarching and ongoing process. Consider the following questions:

- Is the case conceptualization coherent and moving in time through the session (i.e., can I hear and see it throughout the session)? Does it have extension across larger swatches of time (e.g., does it fit for the next session and for where the client and I are headed in the future and according to desired life directions)?
- What is the interpersonal process?
- How are the client and I relating to each other?
- What is the intrapersonal process?
- What is happening inside of me as I work in the session, and is it being functionally brought into the session, if appropriate?

I invite you to explore these questions alone as you listen to the recording or with others in supervision or with colleagues. Notice if there are places where you would like to make a change, growing your therapeutic fluency in ACT through the layers of process.

The Heart of Now and the Door to Next

Hopefully, the client cases presented in this chapter and the Reflective Practice exercises have begun to assist you in thinking about how you can integrate overarching and ongoing, interpersonal, and intrapersonal processes into your ACT work. Notice how the earlier chapters pertaining to your personal growth in ACT are relevant. Awareness of the different layers of process is likely to emerge only through a personal practice. Getting to the heart of ACT involves embedding the six core processes into the larger, ever-changing flow and connection between you and the client. The ability to see into the heart of your client with understanding, kindness, and care by recognizing what flows within you as well as between you and your client is part of ACT work, and means bringing your personal openness, awareness, and engagement into the therapy room.

As you read the upcoming chapters, consider how you can use the material to promote growth of your own mastery of ACT. Consider how these processes inform the heart of the work. Let's turn next in chapter 8 to taking a closer look at the therapist experience in working with a number of clinical issues from the process perspective.

CHAPTER 8

The Clinician's Experience: A Closer Look at the Intrapersonal Process

The best and most beautiful things in the world cannot be seen or even touched. They must be felt with the heart.

—Helen Keller

Effective ACT is revealed through healthy engagement in life. And for most therapists, a fruitful treatment will mean that the client has moved toward healthy engagement in interpersonal functioning, characterized by a sense of connectedness and belonging; toward having a presence to life as it unfolds; and toward routine and active participation in choice making linked to values. Effective ACT isn't found in a hope or dream for the absence of difficulty; instead, it is in the rich participation in life, both its ups and its downs. Indeed, "Healthy and engaged living, from an ACT perspective, includes the full range of human emotion—pain and joy; human thought—that categorized as positive and negative; and human physical experiencing—all bodily sensation" (Walser & McGee-Vincent, in press). This assumes a sense of self that is whole and capable. Engaged living has no arrival; there will always be a choice right up until the end. Suffering will be a part of this process. All humans will encounter their own measure of pain. Therapists will meet their individual level of difficulty both in and out of session.

Skill and facility with ACT, then, will come not only through persistent engagement with written material, workshops, and supervision, but also through the acknowledgment of this shared humanity. In the space of touching the same earth as your client (literally the same therapeutic floor), a process unfolds for each and in between the two. Connecting to your own therapeutic challenges and growing your awareness of the places you get stuck in therapy will provide you the opportunity for change or the possibility of bringing something meaningful into the therapeutic relationship that can be used to shape client behavior and create beneficial interpersonal growth.

Emotional experiencing in session is almost expected by the client. Client tears, anxiety, sadness, disappointment, anger, frustration, and vulnerability as well as joy and happiness are viewed as part of the process, something to be shared. However, these same kinds of experiences are often expected to be tamed when it comes to the therapist. This is, in part, about making sure the therapy doesn't become about the therapist. This is understandable. The job of the therapist is to assist the client in exploring their fears. It is not to open the floodgates of personal pain. Nevertheless, exploring the different levels of process involves therapist emotion. Being aware of and connecting to the contextual, process-oriented layers of ACT includes the emotional experience of the therapist and how it participates in the ongoing, interpersonal, and intrapersonal arc of therapy.

Therapist emotional experience in session not only is to be expected but also can be shared and responded to in ways that are healthy and therapeutic, and it is indeed part of the ACT core competencies. In this chapter, we will explore therapist emotional experience and personal challenges as a means for you to begin to encounter their role in ACT therapeutic process. Although I cannot cover every experience and challenge, as they are perhaps without number, I hope that the examples provided here will speak to the key issues and assist you in considering how they participate in process. You will discover the intrapersonal process by looking at self-disclosure, personal willingness, personal experience (e.g., therapist challenges and difficulties), and positive emotion (e.g., gratitude and love for the client). This is your felt experience, your heart—part of the intended work in modeling and embodying willingness.

Self-Disclosure

Frequently I am asked about the use of personal or therapist self-disclosure in session. In a recent training, an attendee asked with great surprise about the disclosure I had made during a live demonstration with a client. I had worked with the client on the difficulty of making choices in the presence of intense anxiety. I had shared with the client that I had not acted on my values in the past, instead letting anxiety guide my behavior. I had also shared that I had pulled away from an important decision that later I had mixed feelings about, both regretting and feeling relieved about the choice I had made. I had spoken briefly about my racing heart and dizziness when I approached the process of making that choice. The moment between myself and the client in the demonstration was a bit solemn, and I was vulnerable. The client had responded to my disclosure with curiosity and had wondered why I had chosen the way I did. I had openly stated that part of my decision was based on fear and that I wasn't even sure that if I were to go back, I would make a different choice.

The attendee asking about the disclosure noted that he had never done anything like that in therapy and was a bit startled to see me disclose in this fashion. He wondered out loud if he would ever be able to make such a move in therapy and worried that it would be viewed as self-serving and out of the realm of psychotherapy. Other members of the audience chimed in: "Was this too personal?" "Did this make the therapist look weak?" and "What would my colleagues say?" I listened to the concerns and noted the palpable anxiety in the room (i.e., "Is this what it means to do ACT?"). A few clinicians noted their fear. They were wondering about different things. What will the client and perhaps other therapists think of me? How could I ever do this? I would be too nervous. Will the client leave and never come back? Is this unprofessional? This isn't about me—it's about the client.

Exactly. It is about the client. But rather than listening for the patterns and underlying function, many members of the audience got stuck on the content—the content of what I said and their content about the rules, about the dos and don'ts of psychotherapy.

On the one hand, the therapists were noticing their own anxiety and were speaking to a kind of unwillingness. I don't want my client to see this; I don't want to look weak or unprofessional to the client. I don't want to look like I don't have it together. Is it okay to share my emotions and thoughts with such vulnerability? The answer is yes, with this caveat: as long as disclosure is in the service of the client. I explained my disclosure. The client had expressed, in different ways, feelings of loneliness and a sense of being separate or distinct from others. She was also worried about what would happen in her future if she made a particular decision. One decision would mean independence, a definite value, and a genuine desire, but would be challenging and potentially create more feelings of loneliness. A different choice would mean more dependence, creating feelings of safety and stability, but at the cost of some freedoms. Several things were happening at once; it was a complicated values exploration. She was feeling isolated and not part of humanity. As well, she was wanting to make choices that gave her safety and stability in addition to independence and freedom. Her ambivalence about what to do was accompanied by anxiety. She hoped that her feelings would settle down, making a clear choice more likely. She wanted certainty about the future. Assessing what to do next led to my disclosure.

The self-disclosure served multiple purposes. I wanted to convey to the client a sense of belonging, that she is a part of humanity—she is not alone in her suffering. She and I both have pain; we are both humans. The self-disclosure functioned to get the client and me interpersonally connected—to draw us nearer to one another—and to connect her to the everyday struggle of making a complex values-based choice.

I also wanted to model the emotional process of showing up to ambivalence by speaking to and sharing some of the affective qualities of my own experience. Uncertainty is an experience that can be accepted and explored. Lastly, I wanted to convey that life is a process. That even after we make a choice, there are always more choices to be made, whether values consistent or not. The client's curiosity following the disclosure indicated an openness to what I was modeling and sharing. She wondered about her own capacity to feel anxiety and take action. She noted that she had not explored her dilemma in this way, and we left my disclosure and turned back to her experience and struggle.

I could have pursued multiple avenues in working with the client. Nearly any of the six core processes could have been brought to bear individually. Nonetheless, with this single self-disclosure intervention, willingness, present moment (I noted how I was feeling in the moment), and values were all brought into a therapeutic exchange that enhanced the interpersonal process. Timing was important. I did the self-disclosure after observing a pattern: the client noting multiple times her loneliness and isolation, as well as demonstrating a complicated values experience. She wanted independence, and felt isolated and alone. She also wanted safety and stability, something she experienced living with her parents. With independence she risked loneliness; with security and stability she risked dependence. Drawing upon my history, referring to a time when values-based choices were complicated, I was modeling acceptance while exploring the ongoing nature of experience and making choices in life overall.

On the other hand, the therapists in attendance were also worried about breaking the rules: (1) never self-disclose or disclose very little, (2) be sure to look like you have it together, and (3) therapy is always about the client. Let me respond in turn. First, I invite all therapists interested in ACT to break the rule "never self-disclose." Self-disclosure levels the playing field. We are not fundamentally different from those we serve. You and the client are both humans. Join with your clients in your shared humanity; work together to explore and change behavior. As you may recall, self-disclosure in the service of the client is a core competency in ACT. For rule two, see the response to rule one. And for rule three, yes, therapy is about the client, but, as mentioned, not inside of a vacuum. You are there to support, work collaboratively for change, and shape behavior through modeling and contingencies. You are developing an alliance; that alliance takes two people joined to encourage healthy movement forward. Therefore, I will suggest that therapy is about you as well.

Self-disclosure in therapy is, in one sense, easy. Making comments about what you experience and about things that you share in common with the client may flow in a quite conversational manner. Some of what you reveal about yourself will be more or less consequential. In another sense, self-disclosure is more challenging. It has its own subtleties and can be problematic if awareness and function are missing. As well, listing self-disclosure as a core competency for ACT may place it at risk. Disclosure *can* become a rule. In this case, disclosure is for its own sake. This would be problematic and not ACT consistent. So be alert to (a) the purpose of the disclosure to avoid the possibility of it becoming a rule (e.g., I *must* say something about myself) and (b) your authenticity. Let's explore these below.

Consider the Purpose

You can begin to explore the self-disclosure process by asking questions of yourself related to your clients or a particular client. You might ask, *Does this disclosure reinforce or punish a particular behavior or set of behaviors, and does it do so in such fashion as to be consistent with the case conceptualization? Is the disclosure in the service of modeling acceptance of emotion or defusion from mind, and am I doing it at a time that makes sense or at a time that fits with what is happening in the session? Is this disclosure coming from my heart?* If you are unsure about a self-disclosure or you worry that it is more about you than the client, wait. You can always return to the disclosure later after you have had time to digest its purpose.

Many of you reading this section will have already largely "conquered" the self-disclosure competency. You may well be using disclosure in a functional way that is effective and in the service of the client. In considering mastery and heartfelt work, though, I would like to invite you to lean in a bit more. I am not asking for an increase in the amount of self-disclosure you do; I am interested in the kind and depth. There is a place in therapy where we are simultaneously considering ourselves and the client, a kind of "us" consciousness. Inside of this place, I invite you to deepen your awareness of your reactions to your client as well as the impact you have on the client.

Be Authentic

Authentic disclosure in a context of *us* means a larger benefit is possible for you and the client, and beyond. That is, disclosure in the service of helping another to move forward in meaningful ways reaches beyond the therapy room. It can promote change in the individual, change in their relationships, and change in their community. I am reminded that a felt-tipped pen saved space flight *Apollo 11* (Mannion, 2009); small things can have a significant impact.

Authentic disclosure is essential in other ways as well. If you are unaware, then unconscious and small behaviors can have a negative impact. For instance, if the therapist is unwilling and never self-discloses, it has the potential to communicate a number of messages, having an advantage or special authority for instance. A brief look of judgment might communicate lack of importance and value, teaching too much without revealing emotion may communicate superiority, and not being willing to be vulnerable will change the nature and quality of the alliance. Although not likely problematic if these behaviors happen infrequently, should they recur, a cumulative impact is possible. The *us* is damaged and connectedness lost.

Deepening your awareness of your presence and impact, of your intrapersonal experience, will help you to connect more fully to authentic disclosure that supports *us*ness. I should emphasize the value of authenticity. Genuine disclosure is not forced, nor overly emotional. It is not done to *get* to mastery; it is done from a sincere place of serving the client. It comes from the heart, not the head. Consider not only noticing your experience, but taking the perspective of the client, noticing what it might be like to experience you, your words, your revelations, and your actions. Seeing yourself in the therapy room from multiple perspectives: Where do you sit in the room? What does your body language communicate? How are your words received and understood? How are your emotions experienced (or are they experienced) by the client? Is there judgment or distance in your presence? Are you experienced as present to the client? What is your stance and intention, and does it serve the spirit of closeness and connection? What do you truly want for this client? What is your presence while making a disclosure? And, are your disclosures linked to function?

Your willingness matters. It is hard to imagine that from an *us* space, where you are modeling willingness and speaking authentically—whether it be a truth about how you feel when a client behaves a particular way, a story about something from your past, an emotional experience, or a values-based choice—self-disclosure wouldn't be experienced as helpful and part of building a very human connection and alliance in therapy. But if a self-disclosure goes awry, reassess: was it in the service of the client? Adjust. Be open, be aware, and try again.

Manuela's Reflection

Deepening your awareness of your impact on clients feels important. At the same time, therapists have little context in which to see how our presence impacts clients in larger and subtler ways. What would you recommend as a way to work on this?

Robyn: I invite therapists to work on this in a number of ways. Indeed, I encourage readers to engage in this process in supervision and consultation. The Reflective Practice below is a great place to start.

REFLECTIVE Practice 8.1

Consider and write about how your presence and self-disclosure impact your clients. Here are some suggestions:

1. Have a supervisor or colleague role-play with you, and during that time self-disclose. Then have them provide feedback directly related to this issue, including how they experienced you and the self-disclosure you made. You can also have the supervisor or colleague speak to your presence as a therapist more generally, not just about the self-disclosure. Consider their comments and note what you appreciate about the feedback and if there is anything you want to work on.

2. Watch yourself in a video. See what your body is doing. What do you notice when you watch? Share the video with your supervisor or colleagues. Ask others what they notice. Explore both the things you like and the things you would want to change.

3. Lastly, reflect on your experience of others in relation to you. Are they receptive? How do they seem to experience you? Work on self-knowledge and awareness and engage in some thoughtful perspective taking.

As mentioned above, self-disclosure and willingness are intertwined. Addressing personal willingness is an important aspect of the therapist's intrapersonal process to consider as you move toward mastery and the heart of ACT.

Personal Willingness

We covered personal willingness in part 1 of the book, emphasizing its ongoing development. Here we take a closer look at personal willingness as it pertains to both thoughts and emotions experienced by the therapist.

Personal Willingness and Therapist's Mind in Session

One of the more challenging experiences I hear about from therapists learning ACT is conveying abstract ideas and metaphors without getting caught in content. Talking about ACT rather than doing experiential work becomes the norm. There are some fairly standard ideas that most therapists with any therapeutic orientation are aware of about the amount and kind of talking done in therapy. One is the notion of resisting the need to speak after any and every client who makes a comment in group therapy (or individual for that matter). Others include recognizing that silence can be a useful tool for eliciting emotional experience and being aware of the kind of questioning and exploration needed for processing during session. My general thought, however, isn't that therapists are struggling with some of the basic "rules" around therapeutic conversation, but that their struggle is specific to doing ACT.

I would argue that therapists need to be willing to experience longer pauses before responding to a client when in conversation—and this is not the social norm for regular communication. There does tend to be more silence between the parties in psychotherapy in general. Nevertheless, there is a way in which ACT can lead therapists to engage in a very mind-y experience, with the therapist working to deliver metaphors and exercises to the client, and also working to provide an explanation of the metaphors and exercises. As well, the explanation often isn't about "bookending" the experience; rather, it is about providing an actual explanation of the metaphor in concrete terms or repeating an experiential exercise in verbal form. Some of this may have to do with avoidance on the therapist's part, perhaps fearing that the client does not "understand." Some may be about feeling unclear as to whether they can themselves conceptualize why they did a particular metaphor or exercise—so the explanation is about helping the therapist, not necessarily the client. It may be the case that the client is more concrete and truly doesn't understand or connect to the metaphor or exercise. There are other possibilities. Perhaps the therapist is following a protocol and feels pressured to complete all that is detailed on the pages, or perhaps the therapist is generally a "talker" anyway, and their style is just to teach and explain rather than sit inside of silence or confusion. Whatever the reason given, as languaging humans, we are simply prey to our minds. No big revelation there, but we are at increased risk when we are also asked to deliver a therapy that has a different theoretical background from what we are used to and contains "concepts" that can only be experienced, yet we are asked to present them verbally, at least to some degree. Let's take a closer look at some reasons why ACT therapists feel challenged by conveying abstract ideas, metaphors, and exercises without getting caught in content.

First, functional analysis in the behavioral tradition isn't taught in many graduate schools or is presented only in a semester or course. To master ACT, understanding behavior from a functional perspective is essential. Integrating and linking each of the processes to pathological functioning, in rapid fashion, inside of a therapy session, means working to fully appreciate the function of behavior—its antecedents and consequences. This might also mean that, as noted earlier in the book, those with a different training background need to wrap their minds around a new theoretical perspective, grappling more fully with the underlying philosophy of functional contextualism and RFT. However, this process alone is an incredibly heady process. Obtaining the information and skills required to gain a practical understanding of the approach is no small affair. The experiential part of ACT—the openness to emotion, awareness of the moment, and letting go of attachments and control—is often the part of ACT that draws therapists to it. The experiencing process in ACT speaks to the heart and appeals to us as human beings who suffer. Many get connected to ACT via its experiential qualities such as a felt sense of wholeness and acceptance. *But*, this isn't easy to describe. Given the theoretical underpinnings and therapists' desire to communicate about the benefits of experiential knowledge, ACT can lend itself to wordy explanations. It is practical and, frankly, easier to do ACT in that fashion. I would argue, nonetheless, that ACT delivered solely through, or relying too heavily on, words misses its larger goal: holding experience as it is and moving your feet in a values-based direction—verbal and experiential knowledge brought together, interconnected, in the service of healthy functioning.

Second, and at a more micro level, there are ACT concepts that therapists struggle to understand themselves. Take creative hopelessness and self-as-context as examples. Therapists tend to spend a fair amount of time talking about (trying to explain to their clients) these rather than letting the client experience and connect personally to these two processes that are grounded in experiential knowledge. This makes sense; we want our clients to understand. However, as I look across the training I have conducted over the years, I have often noted that therapists are afraid to do creative hopelessness. The fear is about not wanting to make the client feel hopeless. Agreed, and as noted, this is not what creative hopelessness is about. However, once the work of creative hopelessness is understood (it is not about making clients hopeless; it is about undermining control), engaged concerning its full functional purpose, and contacted through experiential work related to perspective taking or other exercises, then there is much less need to explain. That is, if the therapist is willing to sit with what shows up in conducting the work and exercises that are more experiential in nature and that don't particularly lend themselves to explanation, then it allows the client to more readily contact the experience. They don't have to sort through what the therapist is saying in an overly wordy process. If the client can experientially contact letting go of excessive and misapplied control, then a stronger context for its longstanding unworkability is provided. Recognizing that control hasn't worked and is unlikely to work is often difficult, potentially painful, and at times confusing. Sitting with clients as you move together in moments of silence, noticing the paradox of control and the havoc it can wreak, holds its own measure of pain. Willingness on the part of the therapist is essentially a necessity. Otherwise, our own history with control and a personal need to make sure that clients are "okay" (i.e., feel better) can interfere with this process and lead to a very wordy therapy and self-disclosure that is overtly or subtly about making the client feel better.

The case is similar with self-as-context. My personal sense is that this is the ACT process that is likely to be overexplained the most. This makes sense as well. If a client is puzzled or not quite sure what is being explored, then the pull to explain is powerful. Isn't it our job, after all, to help the client understand? Minds are ready to dive right in and work out the dilemma, solve the problem. Even so, self-as-context, consciousness itself, is experienced, felt, and sensed. It is out of the realm of words as an experience, as the words themselves are yet another thing to be observed. Consciousness, observing in the here and now the fluid experience of being, isn't a simple thing to convey. It must be "touched" and cultivated. And the ever-present mind makes this an even more challenging prospect. The eternal commentary draws us in, an infinite grasping hand pulling us out of awareness. We must, therefore, be aware of mind. As noted in chapter 5, mind is like a monkey and its shadow. The mind cannot escape the shadow. The words of Thich Nhat Hanh (1976) are again helpful here, "The mind is like the monkey swinging from branch to branch…in order not to lose sight of the monkey by some sudden movement, we must watch the monkey constantly and even to be one with it" (p. 41).

Taking this brief look at creative hopelessness and self-as-context as examples points to the sometimes subtle and often implicit command to explain when there is confusion—client and therapist confusion alike—or emotional discomfort. Recognizing this pull, assessing its function, and speaking in session accordingly will assist in reducing the mind-y or wordy process that can emerge in ACT sessions. Cultivating mindful awareness, observing the experiencing being and mind, will foster nonattachment and thwart excessive use of words.

REFLECTIVE Practice 8.2

Consider your ACT therapy sessions in general. As you reflect, notice where and when you might tend to overexplain. Are there times when you are more at risk for being too verbal? Are any of these linked to self-disclosure, for instance when particular emotional states are present for you or your client? Also, consider a therapeutic interaction when you have found yourself doing most of the talking in session. What was happening? Was there functional utility to be so verbal? Was there anything that you were avoiding? Depending on what you discover, how might you work on slowing down and talking less in the future?

Personal Willingness and Therapist's Emotion in Session

Therapist willingness to experience emotion in session can play a significant role in modeling and shaping behavior. However, doing ACT is not an emotional free-for-all. There is a distinction between doing an ACT workshop or training and doing ACT therapy. Participants in ACT training often have powerful feelings and may be tearful; they may experience a wide range of emotions based on the experiential work they do and what arises out of it. This has, at times, given participants the impression that intense emotional experiencing is a must, and that good therapy sessions are laden with tears. Indeed, there was a small stretch of time when I felt the need to push the restart button for some therapists after their attendance at an ACT workshop where the message that ACT equals tears seemed to be the takeaway. I am grateful that I haven't needed to push the restart button on this message in quite some time. It is more evident that therapist emotion in session isn't a prescribed tenet of ACT, nor is it proscribed. Rather, therapist emotion should be natural, being true to one's spirit and character in the context of an interpersonal helping relationship. Therapist emotion in session, then, can include the full scope and depth of a feeling being. Language, from my perspective and quite unfortunately, does not capture this fullness. I am taken with the words of Jeffrey Eugenides (2002):

> Emotions, in my experience, aren't covered by single words. I don't believe in "sadness," "joy," or "regret." Maybe the best proof that the language is patriarchal is that it oversimplifies feeling. I'd like to have at my disposal complicated hybrid emotions, Germanic train-car constructions like, say, "the happiness that attends disaster." Or: "the disappointment of sleeping with one's fantasy." I'd like to show how "intimations of mortality brought on by aging family members" connects with "the hatred of mirrors that begins in middle age." I'd like to have a word for "the sadness inspired by failing restaurants" as well as for "the excitement of getting a room with a minibar." I've never had the right words to describe my life, and now that I've entered my story, I need them more than ever. (pg. 245)

Barring this difficulty—language's inability to fully convey emotion—we still find ourselves, as therapists, responding with the full range of feelings available to us as human beings. Our awareness of these experiences and ability to identify them as they occur will inform the therapeutic process and conceptualization as well as the alliance. There is some research suggesting that only 36% of people can accurately identify their emotional experience (Bradberry, 2015). If this research is accurate, then we have work do to. Not just with our clients, but with ourselves. When we are unable to identify our emotional experience, we are more likely to behave in ways that are potentially problematic. It may lead to having less empathy. It may affect how we choose to behave, for instance if we don't know whether we feel guilty or angry. Not knowing places us at risk for acting impulsively or avoiding. It may lead to lumping certain emotions into a single category and thus experiencing a broader sense of emotional distress.

Not being able to identify emotion is an issue not only for our personal relationships at large, but for how we work with emotion in therapy. I return to the importance of self-reflection and mindful observation of your internal experience. Make space in your life to routinely observe and check in. Check in on your heart and your gut. Notice the physiology that is present and the context of your experiences. Are you experiencing an amalgam of multiple challenging emotions or one intense emotion? Are you experiencing joy, or pain, or both in a bittersweet kind of experience? Notice the rise and fall of emotion and which type of feelings tends to be elicited under what circumstances. Being able to identify your own emotion will help you discriminate when and how to bring these experiences into the therapy room.

Whatever your emotional experience that arises, its expression in session should be authentic. This genuine response from you should be linked to its intended function, whether it be to deepen the therapeutic relationship or to model openness to experience or to normalize an emotional response. The intrapersonal process is a vital part of the ACT therapeutic work.

There are a few additional things to consider. When expressing emotion in session, it will be important for you to be aware of the client's well-being. Letting clients know how you feel means sharing a broad range of emotions, anger and frustration included. The method, timing, and tone will be critical. Sharing frustration, for instance, before the client trusts that you are there to help may lead to a therapeutic break in alliance and even dissolution of the therapy. Sharing frustration or even anger (as well as other emotions) should be thoughtfully worded and done in such a way that it can be heard, its context explored. As well, there are times in therapy when you may be experiencing emotions related to personal life events. These might be useful to share with the client, but awareness and self-reflection should be a part of that decision-making process. Consideration of the case conceptualization and the function of disclosure should play into the choice. Only talking about your emotions linked to your own history or circumstances may not be helpful and isn't considered ACT consistent.

Let's explore an example to illustrate further. A few years back I was seeing a client who was trying to decide whether to leave her partner. I had just been through an emotionally challenging divorce myself. Some of the information the client shared with me touched on my personal experience. At times I felt anger in response to what she was saying about her partner, at times a sense of sadness about the prospect of her being lonely in the future. I was quite careful about what I shared. I expressed these emotions but was thoughtful about how and when. I spoke only very briefly of them and couched them in terms of normalizing some of her experience (she wasn't sure she should feel anger) as well as exploring willingness to experience some of what might show up should she choose to leave her partner. The function was to invite curiosity about emotional experience, including difficult experiences that can appear even when choosing to live according to your values. I was conservative given my personal circumstance but not avoidant. I authentically shared my experience as it showed up in the room. I modeled my genuine reaction, but I did not inhabit the therapy with my personal experience. I didn't join her in a dramatic exploration of the difficulties of divorce. Instead, I met the client in a genuine process of willingness, discovery, and engagement in choice.

Given that the client can elicit intense emotions from you, and that self-disclosure is a core therapeutic relationship competency, I also want to speak to the above example to draw the distinction mentioned earlier in this section between workshop experience and therapy experience. If I had been attending an ACT workshop during my divorce and had been invited to explore emotional material, as is often the case in these trainings, then I may have been quite emotional, at times finding it difficult even to speak. Context matters, and the emotional work done in a workshop doesn't mean that this is the type of emotional expression that must be conveyed in session. Again, experiential work does not equal tears, nor does it equal no tears. Instead, experiential work is about the observed and felt. It is about connecting to the fluid experience of thinking, emotion, and what is sensed in the moment. These are tied to broader processes of healthy being, such as creating a stance of acceptance for these events or stepping forward in the face of difficult outcomes. It is about instantiating flexibility or variation in response across time through contact, rather than avoidance of what it means to be an emotional being.

Personal Experience

"Seeing" clearly in the present is also part of recognizing when and how to model and provide feedback to the client based on emotional experience as part of the interpersonal process. Therapists largely do okay in expressing emotions when they are demonstrating empathy for a loss or sadness around a situation. It is still important to recognize, however, that we add our own interpretations to client behavior based on the "haze" of our mind—our personal learning history. It is inside of awareness that we can appreciate and understand these "filters," developing hypotheses and interventions based on case conceptualization and functional analysis rather than on our single reaction or response to the client. This is where patterns of behavior become particularly important. A single emotional response to a client may not be the one to act on (it depends). However, if a pattern of emotional responses is elicited routinely, either in a session or across sessions, it may prove useful to give feedback based on this experience. In what follows, I will explore a couple of the more challenging emotional experiences that therapists encounter in session, noting the importance of willingness in working through these experiences and exploring the "how" of bringing them into session. The discussion isn't comprehensive of all emotions that therapists encounter when interacting with clients. Rather, I selected a few to invite you to consider your own internal experiences, and when and how they show up in session. You might consider asking yourself questions about your own emotional experience when you are considering disclosure: *Am I talking about myself right now and my own needs?* or *Who owns this experience and is it relevant?* I also recognize that there is an interpersonal flow between you and the client, each eliciting emotional response from the other. Be aware that there is a learning history behind the event, not just the emotional response. It is this history that calls for pause, allowing time to consider how you choose to respond to the client.

Feeling Intimidated

From time to time we all encounter a client who behaves in such fashion that we feel intimidated and we experience intimidation's accompanying fear and anxiety. I once worked with a mental health provider in therapy who was well trained in third-wave behavioral therapies. Knowing this made me a bit apprehensive, but that wasn't the intimidating part. What made this client a challenge was her interpersonal style. I will share a bit of a typical dialogue that occurred before I was willing to share with her something important and needed to balance our therapeutic relationship.

Client: I was frustrated again this week with the person I talked about in the clinic. She just makes me crazy. It seems like she goes out of her way to make things difficult. She is disorganized. She comes to meetings late and doesn't guide us through our agenda very well. I am at the point where I don't feel like showing up, but I have to. It is this kind of behavior that I get so twisted up about. It isn't respectful.

Robyn: It makes sense that you would feel frustrated by that.

Client: No. You don't understand. The amount of anxiety I have around this is over the top. I leave feeling angry every week. She should let someone else lead this meeting. I can't possibly be the only who is frustrated by it.

Robyn: Have you talked with others? Shared your experience?

Client: No, why would I do that? They all like this woman.

Robyn: (*cautiously*) Are you experiencing isolation?

Client: (*aggravated*) No. I am simply trapped. I sit in these meetings just trapped. I get more angry and then feel less like I want to be a part of the group.

Robyn: (*cautiously*) Tell me more about this trapped feeling.

Client: (*shaking her head and sitting forward in her seat*) No, that's really not it. I feel out of control and like I am wasting my time and like I can't do anything about it.

Hopefully, you see the pattern—nothing I say is accurate, and she starts every response with a "no." This type of interaction went on for several months. The client was aggravated, and I grew ever more cautious. I became increasingly quiet across time. I was afraid to share my experience of her. I was intimidated. She was a bright, thoughtful woman who came to therapy religiously, yet I couldn't find a way to get it right. I couldn't hear her in such fashion or speak to her in such a way that nearly everything I said wasn't met with disagreement. I thought about this client more than usual outside of the therapy session. I found myself becoming more frustrated and at times even angry at our interaction. I was growing more worried about being ineffective. I also, in truth, was afraid of failure with this client; she being a provider in the community meant she talked with other clinicians and interacted with people who "ran in my circle." Therapy was stuck.

I sought consultation and looked more closely at my experience. I was willing to feel the anxiety and fear that came along with intimidation, but I wasn't speaking to it. More importantly, I wasn't willing to feel the pain of rejection from this client. I wanted her to like me. I wanted her to share with others that I was doing a good job with her in therapy if that happened to be the case. Ultimately, I was getting what I didn't want: I wasn't doing a very good job in therapy, and we were circling around the same issues without much progress. I made a commitment to make a change. What I said and how I said it were important. Merely stating that I was afraid she wouldn't like me and that I was fearful she might tell others that I wasn't doing a good job didn't quite make sense. It didn't address the function of her pattern of "no" behavior that elicited the feelings associated with intimidation and led to my shutting down as her therapist. My wanting to be liked was about me. Pursuing that avenue would have turned into a session about meeting my needs. It wouldn't have been in the service of the client. Instead, testing a hypothesis that this kind of behavior was leading to problems in interpersonal relationships in her broader experience made sense. One of her key complaints was her relationship with her husband.

Robyn: I have been thinking about our therapy sessions and have noticed that we seem to have slowed down. But more importantly, I have been watching my own reactions across time and have seen that I have found myself pulling away, getting quieter.

Client: I noticed too. I have been wondering if this is working. I have been thinking about shifting to every other week.

Robyn: (*rather than responding to the content the client gave*) I've noticed that I have been feeling afraid. And, I have been hesitant to say that I am feeling that way. I notice myself reacting to something you say and feeling intimidated by it. It seems when I respond to something you have said, it is almost always met with a "no." Like I can never get it right.

Client: You sound just like my husband.

Therapy opened back up again. We explored what it was like for me to be met with a "no" every time I responded to her, and she connected to something new. She hadn't realized that perhaps she was eliciting fear and intimidation in her husband. She noted that he tended to be quiet in his interactions with her. She also noted that she found this frustrating with both her husband and me. The quieter it got, the more difficult things became. She spoke about her need to be completely understood, no room for error. She was exacting and wanted others to be the same in reflecting on her experience. She felt anxious and misunderstood when others couldn't accurately reflect her internal experience. We explored the impossibility of this expectation and the bind that it placed her and her husband in—and her and me in. We looked more closely at the function of the "no," noting how it helped her to control her own feelings of fear linked to being misunderstood and the associated feelings of rejection. It became clear that she was getting more of what she was trying to avoid. The more exacting ("no") she became, the quieter those around her became, and the more she felt rejected. She was trapped in an unworkable avoidance pattern. In therapy, we focused on willingness to feel fear and rejection in the service of building relationships.

In this work, it was my job to keep track of my intrapersonal process. Was I responding to my client based on my fears? Or was I responding based on what was in the service of the client? If I recognized a sense of disconnection and a pull to make the client like me, I worked to observe and let go. If I recognized a sense of disconnection and the pull was to comment on frustration or thinking that I was always wrong, then I could say something to the client based on the interpersonal process between us linked to the case conceptualization and her reason for therapy. Awareness of my own experience was essential to the work. Consciously choosing to stay focused on client need and the ongoing conceptualization of the client, as well as the arc of therapy, played a critical role.

Waning Compassion

A therapist once said to me, "There are just some clients who I can't seem to offer another ounce of compassion to, not for their situation, dilemma, or emotional difficulty." We explored this issue a bit, and I shared with her that I didn't believe that compassion was the problem. As she stated, compassion is something offered—an action. It is taking a stance of willingness in the presence of pain. It can be selected, even in the face of difficult emotion. Rather, I thought she was speaking to loss of empathy for clients who routinely elicit disappointment and frustration, and following a closer examination of her experiences, this was indeed what was happening. The therapist was truly relieved, as she had worried she had "broken a therapeutic bone" by "losing compassion" and may not be suited to do therapy.

Loss of empathy for clients can happen when clients are mainly stuck, are challenging, and, no matter what you do, don't seem to make progress across long periods of time. These clients also tend to "try our patience" in other ways. For instance, they may complain about our abilities or criticize our interventions. They may also be quite stuck in passive reception, rarely taking responsible action for their situation. Or they actively seem to sabotage progress more often than not. Under these circumstances, we may come to feel ineffective or burned out. Disappointment and frustration (and even anger) may be part of the emotional experience we find ourselves encountering during sessions with these individuals.

You may want to consider several pieces of information before you make an emotional disclosure under these circumstances. One of the first things to check is whether you and the client are still working on the same agenda. Over time, the agenda of therapy can begin to diverge. You may be working on emotional acceptance and values-based living while the client is still invested in controlling emotional experience. It is good to check. If you are working on something the client isn't working on, then this disconnect may lead to a decrease in empathy. Typically, once the agenda is realigned, frustration and disappointment decrease and empathy returns. Compassion can always be selected as a part of this process.

A second self-check is about considering your broader experience. Are you feeling a lack of empathy with a single client or is it more comprehensive in nature? If the latter seems to be true, you may be experiencing burnout in general, and you may need to take other actions to remedy this issue. This may not be the time to mention your frustration and disappointment, but rather a good time to engage in self-care. Compassion can always be selected as a part of this process as well (for you and your client).

If neither of these two issues is at hand, and disappointment and frustration are authentically arising in session in response to client behavior, then sharing this emotional experience is likely called for, keeping in mind that it may be part of a larger pattern of behavior worth exploring with the client. As noted above, an authentic sharing of this experience in the service of the client works just as well with these emotional experiences. I should note that the only distinction is in what is said. Carefully worded feedback is essential. I mention this as disappointment and frustration often tend to be linked to expectations or conceptualized futures that have not been realized. However, these experiences are not always about the client. They can be about the therapist. Therapist disappointment and frustration can be the result of our own ideas about what should have happened. If the client didn't complain or criticize, if they were not passive or resistant, if they worked harder, then things would look a particular way.

I think it is okay to want things for our clients. We want them to have well-being, love, vitality. It is normal to imagine how things would work out for the client if they would practice acceptance and take action. Clients who are stuck for long periods of time tap into our sense of failure or inadequacy. The question is, should we share this experience? It depends. Yes, if it can be done in such fashion as to temper and observe negative judgments about the client while acknowledging our personal fusion with a conceptualized outcome.

I once worked with a client for three years and progress was largely nil. He and I went around the same issues multiple times. We did creative hopelessness as an exercise and as a process throughout the therapy. He engaged in mindfulness work and did homework assignments. Yet, the presenting problem didn't budge. Nothing changed. He was locked into a story about himself that he responded to with constant self-sabotage, he fervently resisted any alternatives to his one and only solution, and he was genuinely caught in patterns of rigidity. I became more frustrated and disappointed over time. I struggled to have empathy for his situation given his unwillingness to let go of this single desired outcome. I shared my emotional experience with him, but in a thoughtful and ACT-consistent manner: "I've been encountering an emotional struggle in part because I want something different than what you want. I wanted to see more than what you see in your future. And because that isn't happening, I feel myself becoming discouraged, even disappointed. At times frustrated. Inside of these experiences, I feel a pull to shake things up, to act on the frustration and disappointment by raising my voice or being sarcastic. But I recognize the potential judgment behind those actions. So, I am left with disappointment and frustration. I am willing to feel these, but I don't want to. I want you to do something, so I don't have to feel this…I can only imagine how it has been for you."

The client acknowledged feelings of disappointment and frustration as well. He also admitted how he was stuck, noting his own level of resistance to having some life other than the one he insisted on having, but wasn't getting. Through this exploration, we came to the conclusion that it was time to take a break from therapy. And indeed, this was a fine outcome. We shared our desired futures for him, mine one of flexibility and values-based living, his one of being cared for by others in a particular way. There are all kinds of ways to live a life.

Feeling Overwhelmed

Some clients bring a lot of energy or a lot of difficult problems to a session. They might list or "fly" from one issue to the next with barely a breath in between while presenting with intense emotional experience that challenges the capacity to work. Here, the therapist can feel overwhelmed as well, often having a mix of anxiety and fear, perhaps sadness, linked to stress about the number or type of problems, or both, along with worry about where to start and what to do. Therapists in these circumstances might find themselves working very fast, harder than their client, turning to problem solving and trying to avoid feelings of failure.

So what can you do when you find yourself in these places? First, the best thing to do is slow down. Persist at a steady pace while being aware of the emotional flurry that is coming your way and arising within. Second, be mindful that sharing with the client that you feel overwhelmed often communicates that the client is problematic (i.e., *Not even my own therapist can manage my issues*). It carries with it a small bit of judgment that in most cases will not be useful. Speaking to the particular emotions inside being overwhelmed (e.g., "I notice myself experiencing a bit of anxiety myself when you tell me about all that is happening") will likely be the better choice.

REFLECTIVE Practice 8.3

Reflect on your self-disclosures in therapy. Explore your reflections by considering the following questions:

- How often do I self-disclose?
- What is the nature of the self-disclosure?
- Do I link disclosure to functional outcomes for the client?
- Do I find myself hesitating to self-disclose emotional experience in therapy?
- How do I work with experiences such as loss of empathy or feeling disappointed, discouraged, overwhelmed, or intimidated?
- Are there places where I could expand personal willingness to work on appropriate self-disclosure in my sessions?

Find a time when you were feeling particularly reactive to a client. How did you explore this with the client, and is there anything you would like to change as it relates to implementing ACT?

Positive Emotion in Session

It is much easier for therapists to share positive emotions such as gratitude and delight in psychotherapy. These emotions are reinforcing, and our clients often thrive and feel cared for when we share them. But be careful not to offer these when they don't feel authentic. Sharing gratitude and delight that are conjured rather than felt can be just as problematic as sharing a more challenging emotional response out of context or without purpose. These, too, should be linked to function.

It is just as important, if not more, to share joy and love in the interpersonal experience with your client. Laughter and reminiscing, a moment of presence that feels powerful, all can bring joy to the therapy room. Awareness of these experiences is impactful, growing both the alliance and context for powerful work. I have shared with clients that I am delighted, filled with joy, excited, happy, surprised, contented, and proud. There are occasions when a small bit of embarrassment is tied up in expressing these emotions, but I am willing there too. The feeling of love, unfortunately, can pose more of a challenge. It is so heavily laden with social baggage that simply expressing it can lead to numerous conversations or potential difficulties related to how it is received. Even writing about it here feels hard; it can be mistaken for romance. But that is not the kind of love I speak of. Instead, I am referring to agape, or love for humankind, characterized by an unselfish concern for the welfare of others. Communicating agape in your therapeutic stance and interpersonal exchange can be a powerful experience for both you and your client.

REFLECTIVE Practice 8.4

Take a few moments to remember the last time you laughed or experienced joy with a client. Reflect on the session and notice how the two of you were interacting. Notice if the humor and experience of joy were creating flexibility. Then ask yourself:

- Do I use humor in session and often enough?
- Am I a more serious therapist?
- How do I think about humor in session?
- What do I think about being irreverent?
- How might humor and joy support flexibility?
- Do I ever express love, and what do I notice as I think about doing so?

The Heart of Now and the Door to Next

Intrapersonal process and its expression reflect the recognition of client and therapist being connected to the other, part of a larger fabric—we are in this together, we belong to the same cloth. Recognition of when and how to express and explore your personal reactions in therapy is part of growth in moving to mastery and the heart of

ACT. This can be more confronting, but possibly be of greater importance, when working with clients who prove challenging, who are the "worried well," and everything in between. I invite you to fully consider your emotional reactions as a part of self-disclosure in therapy. Your intrapersonal process matters. Your fears, anxieties, and frustrations as well as love and joy are each part of this intrapersonal and interpersonal experience in delivering the six core processes of ACT in an arc supporting personal change.

The intrapersonal process and self-disclosure can lead to rich and effective interaction with the client. We turn next in chapter 9 to the *interaction*—the relational exchange—between you and the client. We will explore the cycle of the work that can be done with challenging client behaviors as examples of working on the interpersonal process more fully.

CHAPTER 9

The Client and Clinician Experience: Exploring Interpersonal Process Through Challenge

Obstacles do not block the path, they are the path.

—Zen proverb

The continual evolution of the therapeutic relationship is tucked inside the overarching and ongoing process of ACT. An awareness of the current of the therapeutic relationship will assist you in knowing which of the six core processes to bring into the stream of the intervention. Attention to the interaction can support a reciprocal influence process that *includes* the therapist's and client's behavior and learning history as they unfold in the context of therapy. The complexity of this interpersonal exchange and its function can be challenged by arising difficulties in the therapy, seeming to place obstacles in the therapeutic path. But as noted in the chapter's opening Zen proverb, obstacles *are* the path. As well, your ACT therapeutic stance (chapter 10) and your intrapersonal experience—both part of the interpersonal process—will influence your presence and response in the face of the common therapeutic obstacles we turn to in this chapter. As you consider these client "obstacles," pay attention to the need for personal awareness and engagement in therapy as interpersonal process is called upon to assist in moving the therapeutic work forward.

Suffering Is the Obstacle

People suffer. As therapists, we see this suffering and often get connected to it in unique ways that families, friends, and coworkers may not, and we are charged with decreasing the fight with human pain. This means that we often see some of the darker sides of psychological difficulties—including those that may challenge our personal understanding of psychology and functioning, our capacity to believe and trust, and our capacity for empathy or willingness to choose compassion. Indeed, suffering comes in many forms and can involve everything from loss of love and meaning, to struggling to understand and connect, to sinking into an abyss of depression and anxiety, to staying in touch with reality. Each of these, when expressed by the client, potentially pushes a personal "button." Indeed, any of these may pose more notable challenges in therapy depending on the situation of therapist and client. Regardless of the unique struggle of any one individual, compassion for their suffering is the ground the therapist should largely inhabit. This can prove demanding at times. We have all met our so-called limits. We have all encountered clients who push us to our "edge" of knowledge or deplete our capacity to be present, maintain empathy, or offer compassion. It is here that our personal work will prove useful. It is helpful to practice self-compassion at these times, as well as seek consultation. Regardless of the challenge the client presents, it is our work to go easy on judgment of the other (as well as of the self)—holding it lightly; accepting, listening, and seeking to understand; and responding appropriately while honoring the humanity of every client, regardless of how much we "like" or approve of the individual and their behavior.

It is sometimes difficult to know what will lead to a specific challenge for a therapist in a single session. Therapists have personal reactions to particular client behavior. Clients can evoke emotional experiences in us that are not always predictable. Sometimes obstacles are surprising. For instance, you may find yourself challenged by a client who has a similar history to your own. An illness, a death, divorce, or trauma—the list seems endless as to what might trigger reactions that make a session more challenging. It is wise to have enough awareness and self-knowledge to know if a particular behavior or pattern of behaviors on the part of the client seems to routinely elicit strong reactions. In these circumstances, seek consultation. I nearly always do myself and find it invaluable. Finally, keep in mind that these obstacles are the suffering, that their path will be helpful as we now turn to meet them.

REFLECTIVE Practice 9.1

Sit with a paper and pen. First, close your eyes, take a few deep breaths, and then shift into your normal breathing...simply settling in. Second, reflect on the times or experiences of "meeting your limits" in therapy. Where do you find that client difficulties bring an emotional challenge for you as a therapist? What behaviors on the part of the client pose personal obstacles for you? Explore these places, perhaps thinking about particular clients. After a few minutes, open your eyes and take some time to write about what you experienced:

- What do I notice?

- Do the obstacles have a theme or pattern?
- How do I typically respond to them?
- Is there anything I would like to do differently?
- How can ACT help with that change?
- Can I see these behaviors as part of the path?

Challenges in Therapy

In this section, I traverse client behavior that can lead to demanding, tiring, or otherwise problematic and challenging therapy. However, just as in chapter 8, my intent is not to cover all possibilities concerning what a therapist might experience or do under the circumstances presented. Instead, the goal is to offer a potential avenue, from an ACT perspective, for using one or more of the six processes, or interpersonal or intrapersonal processes, to approach these issues. As you read, consider more specifically the interpersonal processes involved, noticing if you have found yourself in any of these situations and how they may have impacted the relationship between you and the client. Also notice if any of the ACT processes could assist to overcome the challenge. I cover topics that I have encountered in therapy or that have been brought to me across the years in supervision or consultation. The list is not exhaustive. There are many other unique challenges and situational difficulties, some of which could be a full chapter on their own. Let's begin.

When You Are Working Harder Than Your Client

It is essential to work hard in therapy as the therapist, both in terms of considering what your client needs to move forward and regarding your personal presence and mindfulness during the session. Attending to, supporting, and working with your client is fundamental to good therapeutic work. However, working much harder than your client is probably not going to be useful.

Nevertheless, nearly every therapist I have spoken to has, at one time or another, stated that they are working harder than a client or clients. It seems that there are three main circumstances under which this occurs: (1) when a client is desperate for things to change but isn't taking committed actions, (2) when the client has a set of conceptualizations or expectations about the therapy or therapist and what will happen in the therapy room, and (3) when the therapist feels that they must fix the client or work hard based on their own conceptualizations of therapy or payment for services rather than the function of the client's behavior.

> ## Manuela's Reflection
>
> *I still can remember a phrase Robyn said to me during supervision, and I use it as a mantra: "Give your client her life back." This simple phrase carries a lot of meaning. It is not about me, or what I want; it is about the client. The client is responsible for their life, not me. This gets me back to looking at the client as whole and capable.*

DESPERATE, BUT NO CHANGE

I once worked with a supervisee who reported the following therapeutic challenge. The therapist was working with a woman who was reporting extreme levels of anxiety. Indeed, during the sessions, whenever invited to do mindfulness work, she would begin the exercise, close her eyes tightly, brace herself in her chair, and after about thirty seconds open her eyes and say, "I just can't do this! I just can't!" (she did not have a trauma history). Week after week she would attend session, often shaking and moving around in her chair, but being quite amenable in session, noting that she was willing to do anything to make things better. She was desperate for something to be different. However, with any in-session exercise or out-of-session work, she simply wouldn't engage. The therapist working with her noted that he felt a sense that he needed to "rush in" and do something quickly. He at one time noted that he was "scrambling around," feeling he must make something different. If silence was in the room, even short periods of it, he felt the "need to fill that gap." The client would say, "I can't" to different interventions, and the therapist would shift gears and try something else. The therapist was working much harder than the client: the therapist had an earnest desire to help, while the client was not making a sincere attempt to do something different.

The therapist was being drawn into a felt sense in the room, a desperate struggle to make the anxiety stop. The client's shaking and vulnerable presence added to this pull. Still, it was important for the client to go to work in session. Stepping into acceptance with exposure to emotional experience and a focus on mindfulness processes seemed to be the best path…for both client and therapist. The first order of business, however, was for the therapist to slow down, notice the pull into the desperation, and do nothing but merely sit in awareness of his own desire to scramble. Letting go of the reaction to feelings of desperation and the intense efforts to control what was happening in the room made space for the work of being present to experience to begin. I invited the therapist to persist, again and again returning to willingness to experience the desperation and the anxiety, but not acting on it, modeling for the client the work needed in the room.

This example points to what most often needs to be done first when you end up working harder than your client: observe your own experience, becoming aware of how you might be participating in the process of working harder than your client. Notice if you might be avoiding, not wanting to feel anxiety about failure, slow progress, the client abandoning therapy, not finding the solution, and so on. Move forward using acceptance and defusion processes. Exploring with your client the dilemma you find yourselves in.

I also once had a therapist tell me how much she was working to find information for a client. She wanted to be informed about the different things the client was presenting in therapy and, consequently, was spending a fair number of hours outside of the therapy room doing web searches and reading articles and reports. This is a perfectly fine thing to do, and it is crucial for us as therapists to stay informed. In this case, however, each time the client would come to the session, the therapist would talk about what she had learned and would tell the client how she would pursue more avenues after the session. The client wasn't participating or placing any effort into a similar endeavor. Inviting the client to participate, perhaps even to teach the therapist, is also a perfectly fine thing to do. It supports connection and a leveling of the playing field between client and therapist.

Additionally, homework is the vehicle for generalization and has been shown to significantly improve treatment outcomes (Kazantzis, Deane, & Ronan, 2000; Kazantzis, Whittington, & Dattilio, 2010). Making clients aware of the Kazantzis et al. research as well as linking homework to values-based living can assist with motivational issues. Having an authentic conversation about the possibility for change without action, in and outside of session, may also prove beneficial. Working to create a context of openness, awareness, and engagement includes the therapy session and life outside of the therapy session. Engaging in exercises in the room is undoubtedly part of this process; however, translating that work to broader life experience will be incredibly difficult if the client isn't cultivating behavior change outside of session. Additionally, therapists who continue to work harder than their clients may begin to experience other struggles including feeling angry or disappointed with their clients and even risking the loss of empathy. Therapists don't hold the monopoly on knowledge and work; it is a shared responsibility between therapist and client. Inviting the client to take responsibility may prove difficult for some clinicians or under certain circumstances with clients. Working with your own willingness processes will be helpful here.

STAGNATION INSIDE OF EXPECTATIONS ABOUT THE THERAPY

Some clients will come to therapy with a hope that the therapist will hold a miracle cure, have a perfect answer, or otherwise solve the problem. This client might be naive, overly optimistic, unrealistic, or simply arrogant about what will happen in the session. Educating the client about the process of therapy may assist, but defusion from specific and problematic conceptualizations may also be warranted. Collaborating with the client includes setting the expectations for therapy—that a good outcome will not be based on feeling better necessarily but on increased engagement in life. There may be a number of reactions to this "news." Those who are naive to the process of therapy may well join the therapist in this endeavor. Those who are arrogant about the process of therapy may not.

Let's take the latter case as an example of things to explore when the client's expectations are overwhelming the therapy. The therapist will want to consider if the arrogance being demonstrated in therapy (e.g., I expect you to do certain things and to give me what I want) is linked to a pattern of behaviors that are more generally causing the client difficulties. If yes, then an authentic exploration of these issues as they arise may be part of the therapeutic interpersonal process. This can be challenging, especially if the client is demanding (e.g., I thought you were just supposed to listen to me when I came here) or confrontational (e.g., I pay for this therapy). It is in these circumstances that compassionate persistence may prove useful. Repeatedly and softly confronting the client's expectations and demanding and confrontational behavior may require willingness to feel anxiety and loss (e.g., the client quits therapy) as well as interpersonal feedback. However, simply persisting in soft confrontation because you don't like the client's behavior will generally not get the desired outcome. It may even lead to argumentation and convincing. Rather, persistent yet soft confrontation is designed to promote observation of personal behavior on the part of the client and should be based on their values and the desire to have them come to life. Being persistent can prove challenging for the therapist. I often hear about fears of persisting and its impact. Yet, if done from an open and compassionate place, it can be effective in assisting the client to see that therapy is about opening a process to life, not achieving a positive internal state.

STUCK IN FIXING THE CLIENT

Finally, working harder than your client can be grounded in a genuine desire to help, but it may also be linked to an underlying story that it is the therapist's job to "fix" the client. It is here that you need to observe this desire, noticing how it might pull you into behaving in ways that are unworkable in session. This desire isn't necessarily wrong or bad. The word "therapy" itself means "to serve." We are in service to our clients. This is why it is particularly helpful to ask what is expected by "fix." Staying oriented toward fixing lives rather than fixing thoughts, feelings, and sensations is the work of ACT. However, many therapists have secretly shared with me that they still feel the need to fix in the latter sense of the word. Presence to what arises when you resist this kind of fixing while continuing to serve is your best ally.

Maybe what *is* ultimately problematic is when the therapist takes on the majority of the *responsibility* for achieving change. Consistently working harder than the client, even when it is toward "fixing lives," is not likely to be a successful endeavor and the interpersonal relationship becomes strained. In all cases of working harder than your client, it is essential to stay connected to this notion: change in behavior happens when the client does the work.

Carlton's Reflection

The sentence "This is why it is particularly helpful to ask what is expected by 'fix'" is so simple to read, but I think far from simple to achieve, or possibly to even fully commit to trying to achieve! First, though, maybe the word "fix" needs to be dealt with. Talking about "fixing the client" pulls for an automatic negative reaction in me. "Fixing" concerning people and their psychological content just doesn't sound like the right thing to be doing. It implies people are broken, and it depersonalizes people, reducing them to machines. "I fix folk" is just never going to be a strapline on a therapist's website (I hope not anyway!). However, "thinking and feeling differently" seems not only less problematic to me personally, but also probably what most clients at the beginning of therapy understand the task at hand to be.

In the second wave of cognitive behavioral therapy, therapists talk about "socializing the client into the model," whereas ACT therapists might talk about the parallel endeavor of "creative hopelessness." Both processes function, among other things, to set out some fundamental assumptions on which each therapy is based; they set the scene for what lies ahead. In the case of ACT, this process also involves establishing a transition from the common understanding of the therapeutic objective (having different content) to the ACT-consistent objective (living a different life). However, I think there are many reasons why the objective of different content resurfaces in ACT therapy, including that it is exceptionally pervasive as an idea in many (Western) cultures, and also we know that having different content is both theoretically possible and achievable in practice. No wonder ACT therapists, particularly under the pressure of clients who are desperate for change, end up driving toward that aim.

> ### Manuela's Reflection
>
> *Many students I work with who are learning ACT find it difficult to discriminate between acceptance and problem solving or fixing. When is it the best time to help and do more problem solving or be more directive in session, and when is it time to step back? Could you expand more on this?*
>
> Robyn: *I typically discriminate by considering the function of the behavior and the context in which it is being explored. If the client is avoiding, then I will want to work on acceptance. If the client is having a practical issue, I will work on problem solving. Workability is the litmus test.*

The Talker

Most therapists will encounter clients who fill their sessions with lots of words. To be sure, with some, it is as if the therapist is in the middle of a verbal downpour—the client changing from one subject to another, talking louder and faster, barely letting the therapist get a word in edgewise—and any attempts to interrupt seem to fail. Other clients will let the therapist speak for a short period and then, without missing a beat, pick right back up with the story they were telling, even if the therapist tried to take the session in a different direction. One of the reasons for this type of behavior is in-session avoidance. Verbally slowing down can make room for emotional experience—the very thing the client may be feeling challenged by. Noticing what takes place interpersonally inside of slowing down will be fully part of the work.

Subtler forms of avoidance can also play a role. The client may be thinking that venting or spelling out everything is the solution to their problem. For instance, I had a client who insisted on sharing as many details of his childhood as possible. He wanted to spend the majority of our sessions scrutinizing his history, even commenting when I tried to interrupt or slow things down that it was vital for me to know everything about him. Anything left out might give me a wrong impression or lead me to ask him to do something that wouldn't fit his recovery from his anxieties and fears. His excessive detail was the way that I would understand him and therefore know what to do. When he discovered that knowing every corner of his history wasn't necessary to move forward in his life, and that the here-and-now process between us could help, he found himself pleasantly surprised. Other reasons for excessive talk might include intense fusion (see "Sitting with a Great Big Giant Mind" below), focusing on the past or future, and buying into a particular sense of oneself.

Most of us have been taught that it is impolite to interrupt others while speaking, and for the therapist, there is an extra layer of social pressure as it might also be considered unprofessional (i.e., not doing a good job of listening). Warm and attentive listening is built into the role of the therapist, and for many, the idea of interrupting a client brings fear of judgment and worry about the interpersonal relationship. A fair number of therapists have been surprised when they have observed me interrupting a client (including multiple times in a short period), and many have asked if it is okay, noting a fear that the client may not return to therapy. In the case of the talker, however, interrupting will change the context, allowing you to address the function. It opens the door for assessing, exploring, and reducing avoidance and fusion—a necessary avenue for change from the ACT perspective. This also requires personal willingness to experience the discomfort that arises when breaking a social and interpersonal rule while also defusing from the rule itself.

Interrupting can be direct: "Let me interrupt you." This will often give you a little space to say what you need. Be aware of the social rules and go ahead and break them anyway. Being socially polite or recognized as a good listener will not serve forward movement in the therapy in these cases. Inside of this interruption you may comment on the pace and amount of content, directly taking time to explore the function of the talk with the client. This exploration may involve inviting pauses, slowing the talking down, and sitting in silence in the future to see what shows up emotionally in the empty spaces in between the verbal expression. Sometimes even a few seconds can make a difference. It may also involve speaking about how the interruptions are experienced—taking notice of and exploring the interpersonal impact.

The client mentioned earlier who could seem to participate in mindfulness for only a few seconds was also unwilling to bear silence. At one point in therapy, the therapist asked the client to merely pause and let silence be in the room, to notice the experience gently. He sat for about four to five seconds and then proceeded with the session. She eagerly joined back into the discussion and went on with her stories. The therapist was quickly swept away, back to the verbal river, later wondering what might have happened had he waited longer. He noted that he could sense the intensity of the silence, that even five seconds felt immense. He broke the silence not just for her relief, but also for his own. In these cases, I often recommend that the therapist, for as long as they are planning to wait in silence, expect to wait a bit more—the relationship can handle it. Judging time in these situations, unless watching a clock, can be difficult. When sitting in silence, it seems, we tend to overestimate its length. Allowing the space for emotion to be present and experienced, as well as practicing noticing the urges to talk over the top of emotion, is helpful to promoting a stance of present-moment acceptance, and will mostly take more than a few seconds (although there are cases where starting with a few seconds is needed, but graduating to more extended periods of time is the goal). Let the client know that you will need to interrupt in the future as well.

In addition to direct interruption, you can use other methods to break the pace and amount of verbal expression. You might consider getting very quiet and sitting back in your chair. The client may sense the difference between the way you are behaving and the way they are behaving and slow their pace, or ask you a question. You can model pace and content in your response. You may also choose to use other forms of body language such as holding up your hand in the "stop" motion. Depending on the relationship you have with the client and your interpersonal-process skill level, you might also consider other, more irreverent forms of indicating the problem in the room.

For example, after multiple and varied attempts to help a client be aware of her excessive talking and avoidance behavior, I once yawned and tapped my mouth with an open palm to indicate boredom with the ongoing storytelling. It did lead the client to pause. She stated, "I know this is boring, but I just can't stand it" and then started to cry, the first tears since she had started therapy. I immediately sat forward and became alert and interested, inviting her to stay present to the tears and the experience of feeling this sadness and distress. It was one of the more helpful sessions. However, if you plan to use irreverence, you want to be sure that it is not coming from a place of anger or sarcasm. My yawn was a genuine attempt to communicate something in the service of finding a tiny crack, making just enough space to explore the possibilities of willingness over control. My purpose was clear, linked to the case conceptualization, and ensconced in the accepting, present, and heartfelt therapeutic stance (see chapter 10). I was also aware of what I would do next in the session if the yawn caught the client's attention—pivot to present-moment awareness and the impact and purpose of excessive talking from a place connected to a genuine desire to help the client move forward. Stretching, standing, looking bored, humor, and other forms of behavior might serve this same function. Use these wisely.

Sitting with a Great Big Giant Mind

Personally, one of the more challenging client behaviors I encounter is what I call "sitting with a great big giant mind," or profound and pervasive fusion. The experience is one of sitting with a mind, rather than a being (the interpersonal process already challenged). This behavior can also look similar to that of "the talker"; however, its incessant nature is not in the pace and amount of verbal behavior, but in the type. It includes analysis, examination, investigation, and intense desire to understand or make sense of the struggle, and no amount of creative hopelessness or other forms of control as the problem seem to pry the mind apart from its owner. Even attempts at mindfulness fail.

I have only had a few of these types of clients in my practice over the years but have also heard from other therapists about this issue. It may be that the ability to fuse and defuse is on a continuum, and some individuals are at the far end of fusion. Working around or through this issue poses a challenge. Let me share an example. I once worked with a female client who came to see me at age seventy-four. She was having fears related to breathing. She suffered from sleep apnea and worried excessively about her CPAP (continuous positive airway pressure) machine, although no defects or problems were found with the machine or the way she was using it. Her presentation had obsessive qualities, but she did not meet criteria for obsessive-compulsive disorder. She had been to many medical doctors, tried multiple CPAP machines, called CPAP companies, been for special fittings, worked with CPAP specialists, talked with her husband, talked with her daughter, seen multiple therapists, tried antidepressants and various psychiatrists, and none could seem to fix her "CPAP problem." In the time that we worked together, we were also able to discover that she was concerned about dementia and death, although she had no signs of dementia and was in relatively good physical health (she ate well, exercised, and had no other chronic illnesses). Regardless, I was of the opinion that dementia and death were creating a great deal of existential anxiety. As we worked together around her fears and worries, we repeatedly landed in the same place: if there were just a way to be sure her CPAP would work properly. Every exercise, every metaphor, every bit of processing was met with a return to trying to figure this out, understanding it and analyzing it from all angles. She figured something should provide the answer. The brief dialogue below illustrates the issue and reveals one form of the *big mind* dilemma. This exchange occurred in about session 8 after we had been working from an ACT perspective since the beginning; much effort on mindfulness and acceptance processes concerning her worries, death, and dementia had been spent.

Robyn: Have you noticed that we have visited this place many times in therapy? We have explored every corner of this problem and find ourselves in the same place. It is like we make some movement and then get stuck, but we get stuck back where we started.

Client: I know. I was really thinking about calling this new company I found online to see if they might be able to help me with the CPAP machine.

Robyn: Let me see if we can really try to catch this again. You say that you were "really thinking." Notice that your mind is taking you down a familiar path. It's trying a new solution to fixing your worries about the machine…

Client: Yeah, I know, I planned to call them, but I thought I might just get the same answer I always get. I would find myself getting frustrated, getting put on hold, getting the runaround, and then being disappointed.

Robyn: I am going to try to stick with this and see where we land. You said, "I *thought*" and then you said some other things about what happens when you try this strategy. I want to recognize that you had a thought. And that thought was part of a broader set of thoughts about this issue. Your mind is very busy with this, and I am hoping we can step back and see that your mind does this, that it gets you stuck in trying to solve the unsolvable. Do you get a sense of what I am saying?

Client: (*sighs*) Yes. Maybe I am just not being smart enough about it.

Many conversations and attempts to defuse went on in this fashion. My frustration grew and my hope for something to be different waned. In session 9 we returned to a focus on values-based living, explored meaning and time in the context of her age, and ultimately decided that the sessions were coming to an end. I wish I could report that we were finally able to connect with defusion, exploring the space between her and her worries, her and her mind, hoping to get some freedom from the effort to problem solve the "issues" with her CPAP machine and turn toward living. Regrettably, we didn't end in that place, and even in the final session, she noted that she would be working to figure out how to fix her CPAP problem.

I made other efforts to assist with defusion. I tried to physicalize her mind on multiple occasions, completed the Titchner exercise (see Hayes et al., 2012), took her "mind for a walk," and routinely related to her mind as a separate entity, but each time she came back with similar responses to the ones above. As I reflect on this client, I find that I would do things differently and work to introduce a fair bit more perspective taking and worry less about defusion, focusing more on the impact of the impasse on our interpersonal interactions. Pivoting to existential issues appeared to hold some impact, getting present to this issue from an interpersonal space may have proved more effective. But the mind that sat before me was rigidly set on the CPAP issue.

There are ways in which this example, to me, reflects a kind of built-in effort to remain innocent or ignorant—a quality of sadness and desperation to avoid death. To a degree, I can touch from a heartfelt place why this client work seemed to fall short when I connect to the pain of coming closer to the end of one's life. Avoidance in this area can be powerful. I occasionally wonder what the therapy would have been like if I had more earnestly shared my internal experience and thoughts about coming ever nearer to one's end. I remember hesitating, and now think that I would not.

There is another kind of "great big giant mind in the room" quality of thought that has less innocence. This kind of big mind is about knowledge, being smart or perhaps clever. It includes behavior that is about working to figure it out because analyzing and understanding is part of something interesting related to philosophizing, is part of making sure others don't judge you as "stupid," is an effort to outwit your therapist, is part of a desperate attempt to control, or is born inside of narcissism and excessive self-focus. It is here that two main things seem necessary to do. First, a fair bit of time will need to be spent, during and across sessions, in powerful and persistent creative hopelessness and control as the problem work. Leaning in with tenacity will be part of this process—pointing again and again to understanding, cleverness, and figuring it out as a part of the problem. The tenacity, although firm, should remain inside of a compassionate ACT therapeutic stance.

Second, connecting the individual to experience—an experiential way of knowing the world—will be paramount: supporting the client in present-moment work that focuses on noticing analyzing, but also becoming aware of the five senses and all that human beings experience (not just mind). Mindfulness work and returning to emotion will be the better part of the therapy journey. This can be challenging for the therapist, as ACT can pull for mind-y explanations and clever answers. As a therapy, I think, its philosophy is not only interesting, but fun to entertain as well. Be mindful of your own desire to turn to philosophy, smart answers, or witty techniques in these circumstances. Letting the ongoing flow of experience, not as a conversation, but as itself, into the sessions will be of great assistance in working with a great big giant mind and can be borne out in the interpersonally experienced process.

The Sound of Silence

Silence, as touched on earlier, can be a powerful part of therapy, lending itself to the bubbling up of emotion, the observation of experience, and mindful interventions. For much therapeutic work, it can be useful and welcome. From time to time, however, silence in therapy is challenging, primarily if it is initiated by the client. For example, a therapist once sought consultation about a client who would come to the session and sit silently for the full hour. Week after week the therapist would try different approaches: letting the silence be, asking questions (which would typically receive a shrug of the shoulders), or talking about issues relevant to being in therapy. This kind of behavioral shutdown can make it quite difficult to do much therapy at all and can stymie the therapist concerning planning a way forward. The therapist seeing this client also noted that she felt like she was "stealing" money from the client because not much was happening in the session. She complained of experiencing guilt and wondered if she should refer the client on to a different therapist, someone who could break through the silence. She felt ineffective and questioned her ability, at times feeling such frustration that she would harbor anger at the client and hope that the client wouldn't show up to the next session. As can be seen by this example, silence experienced in this fashion can be emotionally difficult for the therapist, creating even more distance in the interpersonal experience. Our minds work to solve the problem, and our emotions can range from frustration to anger to despair.

Many of the usual approaches to helping the silent client engage, such as asking questions, waiting silently with the client, and providing a holding and warm context, are worth trying. Nevertheless, willingness on the part of the therapist is potentially the most valuable process. Being patient while observing your emotions, thoughts, and sensations, resisting forcing the client, is possibly the more productive path. Take time to sit in the silence as well. It is not your enemy. Note the experiences that arise within you, paying attention to the judgments, evaluations, and wonderings that you have about yourself, your client, and the situation between the two of you. Take time to consider the function of the silence (noting your intrapersonal experience will be helpful here as well). Is it straightforward avoidance, or more subtle forms such as being right or waiting to feel safe and in control? Is it fear about the future or anger about others? Gently testing these hypotheses with the client may assist. Commenting and exploring the space between the two of you inside of this silence may prove fruitful. You might note to the client an authentic interpersonal experience that speaks to process: "I feel as if we are miles apart from each other and I have a desire to close the gap," or "This silence is simultaneously welcome and unwelcome; I want to hold it and run away from it at the same time." You might also note to the client the interpersonal impact, doing so with caution: "The silence leads me to wonder about whether I can help; I feel anxious about it. I am wondering if you have similar feelings or wonderings: 'Can I be helped?' 'Is my anxiety too big?'"

In the end, patience, not as a form of merely waiting without complaint, but as a virtue, as a values-based action, will be part of your work in this situation. This is why I am emphasizing willingness, as patience includes waiting without complaint, but also in the presence of discomfort.

When Nothing Is Good Enough

Occasionally, therapists encounter clients who respond to suggestions, feedback, comments, ideas, musings, reflections, and requests with comments that indicate, in some fashion, that they, the therapist, have missed the point. Whatever is said by the therapist is somehow short of expectations or wholly fails. The overall sense is that nothing you do or say will work, nothing will be, nor is, good enough. Clients who engage in this behavior can lead therapists to feel frustrated, impotent, intimidated, criticized, or all of the above. In this context, I have seen therapists shut down—barely speaking during sessions or making comments that are generally about agreeing with the client. Our clients shape our behavior as well.

The usual suspects for the function of this behavior are avoidance, fusion, and being rigidly caught in an identity. For instance, the client might be particularly defensive, feeling as if anything noted by the therapist is somehow a critical comment on their ability, capacity, or desire to get better. The client may be so well defended that nothing will do, not wanting to risk the pain of imperfection or perhaps the implication of not personally knowing the solution. They may merely be fused with the idea that nothing will make a difference, rigidly clinging to a way of responding that casts doubt on any potential path forward. They may be tightly clinging to a notion about their self—heavily involved in a story about why they are the way they are, stubbornly rejecting any movement away from that story as it brings a whole new set of problems. I once worked with a client who had both trauma and moral injury during wartime. He experienced a high level of anger about his past and genuine upset about some of the things he had been commanded to do. He saw himself as a victim. He clung so tightly to this sense of himself that any comments, suggestions, musings, or ideas threatened this identity. Any hint of work that might suggest that he was more than a victim simply would not do. Nothing was good enough to heal his pain.

Therapeutic work that includes the processes of acceptance, defusion, and self-as-context may be useful, but not if you start with "nothing is good enough" as content. If openness to experience, seeing the ongoing process of thinking, and connecting to self-as-context are quickly batted away, it can make the use of what might seem like the antidotes to avoidance, fusion, and overidentification look like little more than a Band-Aid. In these instances, present-moment work that focuses on intrapersonal and interpersonal processes may be the best place to engage the client. Giving voice to your internal experience and the way the rejections impact the therapeutic relationship in the moment as they occur brings the immediate and direct contingencies into the room. It is much more difficult for a client to reject an internal experience you are having in the here and now. It brings awareness to the process of rejection (i.e., what happens inside of this relationship when nothing is good enough) and allows behavior to be shaped directly.

The following brief client–therapist dialogue demonstrates how present-moment work can facilitate the difficulty of trying to get past "nothing works." Let's return to the client with trauma and moral injury that led to a place of overidentification with victimhood. Note how the therapist in the first vignette gets stuck. Also note, though, that the therapist's conceptualization of what is happening for the client and a possible way to intervene is accurate regarding the ACT model. Working with the client to connect to self-as-context in the situation of being attached to an identity makes sense and is a reasonable way to proceed—at some point. The behavior that needs to be addressed is the "nothing is good enough" behavior. Imagine as we come into this client–therapist exchange that the therapist has been working to move into self-as-context work for a bit of time and the dialogue has proceeded in a similar fashion across multiple sessions, although with different content.

Therapist: I want to be sure that I do two things in what I say next. First, I want to recognize fully the pain of what happened to you. I think your wartime experience was difficult and life changing. Second, I want to help you move to a place where you can be more than your past, where you can step into a place where you were a victim, but you don't have to continue to live as a victim.

Client: I will always be a victim. There is nothing that will ever change that… there is no place to step into. These experiences have invaded and ruined everything in my life. There is nothing anyone can do.

Therapist: (*working to come back to empathizing with the pain of being a victim, fearing the client did not feel empathized with*) Yes, it is the case that what happened has had a tremendous impact. That is why I wanted to be sure that I noted two things in what I said. I really hear the pain of what happened and want to validate and be present to that.

Client: (*with a bit of sarcasm*) Tremendous impact doesn't begin to capture what I have had to live with…people just don't get what it is like to be in war—(*getting more animated*) they just don't get how hard it is!

Therapist: I can't imagine the difficulty…(*silence*)…But I do think there may be a way forward. A true possibility that is about connecting to something beyond what happened to you during that time.

Client: If you had a way forward, it would be a miracle, and there are no miracles.

Therapist: I agree, there are no miracles, but perhaps there are alternatives. Perhaps we can do something here that can shift this if you are willing…we can get a different perspective on who you are…

Client: (*interrupting, emphatically*) I have tried every alternative (*falls silent*).

Therapist: (*empathizing again*) This really is painful.

Client: You don't know the half of it.

Therapist: (*quietly*) You're right. I don't.

Client: I'm not trying to be right; it is just the way it is.

Here, I hope you can get a feel for what is happening in this session. The therapist is trying to do self-as-context work with a touch of willingness, while making sure to validate the pain the client has experienced, but it is being met with "nothing is good enough" behavior. I wouldn't be surprised if the therapist in this scenario felt quite stuck and unsure about what to do. You can imagine that anything the therapist says will be met with something indicating how the therapist has fallen short and is failing to understand what it means to be a victim of this sort. Nothing said will capture the experience of the pain nor allow for movement forward. Let's now pick up this same dialogue but pivot to present-moment work. Note that this pivot won't always have the intended impact or be the precise thing to do in a session. However, the capacity to pivot into a different process based on the function of the behavior, an awareness of your intrapersonal experience, and the interpersonal impact is the critical point. Let's pick up in the middle of this exchange.

Client: (*with a bit of sarcasm*) Tremendous impact doesn't begin to capture what I have had to live with…people just don't get what it is like to be in war—(*getting more animated*) they just don't get how hard it is!

Therapist: I can't imagine the difficulty…(*silence*)…But I do think there may be a way forward. A real possibility that is about connecting to something beyond what happened to you during that time.

Client: If you had a way forward, it would be a miracle, and there are no miracles.

Therapist: (*with sincerity*) I am noticing a tightness in the muscles of my neck.

Note here how it would be challenging for the client to respond with a rejection. It is about the therapist's experience and is short in nature, giving less content for the client to push off from. Making this kind of statement should be authentic, true to the experience of the moment. It is not intended to be a gimmick but can function to shift the context of the current experience.

Client: (*perplexed*) Yeah…this stuff gives me tightness in my neck too.

Therapist: Perhaps we can connect around that experience…because…because we don't seem to connect around any other.

Client: This is what this stuff does to you: makes everyone feel like they can't connect with you. It is the story of my life.

Therapist: (*doesn't follow the content of what the client said but stays with present-moment interpersonal work*) Let's feel this tightness and look to see what is happening between us. (*pauses for a moment*) What I am noticing is that I feel rejected, like nothing I do or say will matter. It is like a wedge is between us.

Client: There isn't a wedge. I am here, aren't I?

Therapist: There it is again, tightness…rejection.

In this latter exploration, the therapist can work with the client to explore what is happening in the interpersonal space between the two of them. The challenge of nothing being good enough is brought into the experience. Both can work together on what this means for the therapeutic relationship, the movement of therapy, and, ultimately, the outcome they are each hoping for—freedom from the cost of being a victim. In place of the therapist's being shut down, space for some kind of movement is possible. The therapist may still have feelings of frustration, impotence, or intimidation, or they may still feel criticized, but these can be explored directly. Noting their experience in the moment and how it might be interfering with healthy change may help the client to be aware of their behavior, potentially leading to efforts to shift away from the stagnation it can cause. Present-moment process brought into the experience would, hopefully, also free up the therapist to do the much-needed additional work linked to their conceptualization of the client struggle.

Taking the Poison: Being Right

It is not unusual to experience issues of right and wrong in therapy. I wouldn't be surprised if they could be found in every clinical case. From an ACT perspective, they can be worked through in a typical manner using acceptance and values-based interventions, positing and exploring with the client notions of wanting to be right versus wanting to live. Other ACT books have explored what it means to give up being right concerning emotional experience, letting go, and perhaps engaging in forgiveness as an activity (see Walser & Westrup, 2007; Hayes et al., 2012). I will delve only briefly into this issue as it pertains to the stickier behaviors of indignation, self-righteousness, and holding the world responsible for personal troubles.

Clients can get stuck in anger and annoyance involving issues of fairness and responsibility. They hold out—standing in the position of being right as a way to demand justice or claim that others are responsible for lasting or current problems. One of the critical issues of living inside of being right, however, is its cost. Holding longstanding anger at others is personally costly. It is, as has been said in other ACT texts, like taking the poison and waiting for the other guy to die. Anger often impacts the one who is angry whether or not it affects the individual or individuals the person is mad at. This is not to say that clients who are stuck in issues of right and wrong don't have reason to be in this space. There is often an unfairness, a trauma, a painful childhood at the hands of a parent, and so on that precedes the current difficulty. These should be acknowledged and validated, compassion for a difficult past brought to bear.

Nevertheless, we need to recognize that not only does the world bring itself to us—there is no escape from suffering—we also bring ourselves to the world (and to the interpersonal exchange). And what we do and how we live in the world matter; our personal choices give meaning to our existence. So, in addition to showing up to the experience of sadness underneath the anger, letting go of the past and histories that cannot be changed, and working on giving what came before (giving back what was there before the harm) in acts of forgiveness may be essential to emphasize choice and responsibility (i.e., ability to respond). When I have worked with clients who seem mainly stuck in being right, handing choice about the issue fully back to them is sometimes the only thing to do, letting the consequences of the choice be felt and either heeded or felt again. It may also mean sitting with sadness about a client's choice. In the example below, the client has a history of anger at others for not appreciating his personal views or takes on experience.

Client: I want to tell you what happened to me. I am so angry. I wrote a letter to my teacher telling her off and letting her know that I would never return to her class.

Therapist: What happened?

Client: We were asked to make a comment about why we did meditation. I waited to go last. I didn't think that the other people who were there really connected as deeply as I did to what was being asked of us. I took my time and then decided to read a poem about meditation and then the teacher looked right at me and said, "blah, blah, blah." I was so angry and humiliated. I will never go back there.

Therapist: This is one of your few places of reprieve and practice. I worry about this decision.

Client: It was humiliating. I will never talk to them again, and I wrote a letter saying so.

Therapist: I hear the pain of it, but I also fear the cost.

In this session, the client was seeking validation concerning how humiliating the experience had been. The therapist, recognizing the function of the story (based on a history of a similar pattern of behavior)—to be supported in being right—sidestepped the validation and spoke to the cost of being right. The choice to sever the relationship with the teacher and the center for meditation wasn't necessarily painful or punishing to either. Instead, the cost lay with the client. He cut himself off from a stable place for him to go, a place that he had been visiting for years, a place where he found he could feel less alone and engage in a helpful activity. Even after pointing to the cost, the client remained inside of being right. The intense exploration of how this action was cutting him off from his spiritual work and connection was ignored. He had been humiliated, and he would "show" them the depth of this unfairness by never going back. The therapist handed the choice, with honest conviction, "back" to the client: "You can absolutely choose never to return. This choice is yours, not mine, to make. I honor your choice." This kind of process in therapy, a full and untethered acknowledgment of the client's choice, in a deep sense, hits right at the heart of mattering, right at the heart of honoring the meaning our clients and we create. Taking our own responsibility in the interpersonal process is part of this endeavor.

Challenges to Take Off the Clinical Table

Some behaviors in therapy challenge forward progress. These include (a) suicidal behavior and (b) getting stuck in history and other untenable experiences, such as just wanting the pain to stop. Each of these behaviors calls for some form of problem solving. Suicidal behavior calls for immediate action, safety planning, and assessment of the need for hospitalization. Being stuck in history calls for analysis and understanding, solving the problem of why things are unfolding the way they are. Getting stuck in other untenable experiences calls for imposing some kind of problem solving that assists control of internal and external events (e.g., "Make my partner do what I want"; "I will do what you tell me to do."). Barring taking the appropriate action according to your local laws and ethics when it comes to suicidal behavior, you can set aside problem solving in each of these cases, clearing a path for more useful work in therapy.

Suicidal Behavior

The immediate pull with suicidal behavior is to figure out how to make it stop. It causes a great deal of distress for client and therapist alike. Therapists often feel compelled to act immediately. I agree with this feeling, but what action to take is critical. Rather than speed up and work furiously to resolve the suicidal ideation and behavior, it can be imperative to slow down, setting problem solving aside, leaving it for later in the session. In the moments that you learn about suicidal ideation, bring your full attention to the human being and the pain they are experiencing in the moment. Notice your internal experience; notice what is happening between you and the client in the room. Rapid problem solving to resolve personal distress can project anxiety and fear and may even function to scare the client further, sinking them more deeply into fusion with ending the pain. Letting gentle compassion emerge that communicates that even the most painful of experiences can be held here is different from moving directly to a safety plan. Depending on the seriousness of the suicidal threat, a safety plan can be developed, but not in reactionary fashion—an oft taken path due to fear of suicide. Instead, let your heart and warmth be in the room, guiding the client to hope for a better life. I have stated something similar to the following in session: "Therapy is about life, in here…our work is about engaging this weird, wild, painful yet lovely thing called life. If we make this work about trying to get you to not commit suicide it will take on a very different tone and flavor. It will become about hospitalize or don't hospitalize. Call the police, don't call the police. I hope to make it about something much more than that, something more vital, connected, and loving. Will you join me?" Working with the client, inviting them to join you in flexible engagement in experience and life, will look entirely different from stopping suicide: one filled with the work of cultivating conscious living, the other filled with safety plans. I don't intend a cynicism when focusing on the problem-solving aspects of suicide; I take this part of working with suicidal behavior seriously. But I always want to slow down and see if life is an option first.

Addressing suicide in a more flexible fashion might take different forms in session. In the below example, the client started this conversation in a very casual way, almost as if it was a boring part of a long story. In the first option, we explore what happened. In option 2, we will explore the "slow down" alternative.

Option 1

Client: (*casually*) Well, things are not going too bad…actually…you know, a couple of hours ago I put a cord around my neck and fell back and let it take my weight. I was just getting to the point where I was just about to pass out and I pulled it off.

Therapist: (*alarmed*) This is very serious. That is a significant attempt. Let's look at this. How likely are you to do this again? (*The therapist goes on to check for safety and create a plan.*)

Option 2

Client: (*casually*) Well, things are not going too bad…actually…you know, a couple of hours ago I put a cord around my neck and fell back and let it take my weight. I was just getting to the point where I was just about to pass out and I pulled it off.

Therapist: (*alarmed, but consciously slows down, pauses, lets a little silence hang in the room*) I feel jarred right now, pulled to rush in (*longer pause*). I am baffled about why you might say something like this just as if you were telling me about something as mundane as brushing your teeth. I can feel the giving up inside of it.

Option 2 connects to multiple processes and isn't focused on simple problem solving. It brings in intrapersonal (feeling jarred), interpersonal (pulled to rush in and save the client), and overarching processes (there is a sense of giving up, something that has been brewing across time). This option feels more connected to the moment and attempts to empathize with the client's state. There are many ways to respond to suicide in session; this example points to a multilevel process that keeps the door open for ACT-consistent work (i.e., recognizing thoughts and feelings for what they are, acceptance versus avoidance). Immediate problem solving in this area can shut ACT-consistent work down. The focus becomes about stopping what is happening versus noticing.

Sometimes suicidal behavior is chronic and the therapist needs to simply find a way to move it out of the forefront of therapy. This might include arranging plans that involve self-hospitalization and consequences for suicidal gestures. It might also involve asking the client to take suicide "off the table" for a period of time. Choosing to commit suicide is always an option, meaning it is always a thing that that client can turn to, as much as we may not want them to do so. Taking if off the table, for now, can make space for other work to be done. This doesn't mean it will be ignored if it continues to be present or worsens; addressing these circumstances will be essential. I am arguing, however, that spending session after session working on not committing suicide means that other things are not being explored. Inviting a reprieve can sometimes be useful in this area. Work on acceptance of emotion and mindfully observing painful thoughts can then be more fully discovered, freeing the client from the need to act on suicidal thinking.

Getting Stuck in History and Other Untenable Experiences

Clients can get stuck in hope and desire. These desires and hopes are often linked to forgetting about or somehow miraculously undoing history, changing other people's behavior, or being told what to do so that everything turns out just right. Aside from the creative hopelessness work that these issues call for (e.g., no other history will ever be, I can only work with the person in front of me, and it is ultimately a matter of values-based versus control-based living), it might also be helpful to point to the hopes and desires themselves as part of the problem. That is, a client may hope that you can somehow undo their history, and you can focus on the unworkability of this hope. But you can also work on hope itself. Working with the client to acknowledge the areas of control that appear in a "first level" sense (e.g., undoing history) may be useful, but if you are stuck or it seems appropriate, it may be helpful to look a layer deeper, undermining a kind of "second level" area of control—our deeper desires, our deeper hopes. Letting go at this deeper level brings its own pain. Fully recognizing the problems of hope and desire brings with it deep sadness. We will never have what we have conceptualized, and recognizing this experience is there for both of you may be part of a powerful interpersonal exchange. No matter what hope or desire is fulfilled, some piece of pain will be its companion. Whether it be different from what we thought, or the end of the experience, some form of emotional angst will emerge, especially when we are attached, which is the very essence of hopes and desires. However, the other experience that may occur when letting go at this deeper level is freedom. It is the flexibility to live now, letting each moment unfold as it unfolds—for both you and the client.

> ## Manuela's Reflection
>
> *When I read about difficult clients and ways to work with them and first started working with ACT, I found it was easy to get hooked into the idea of a miraculous move that will make things better. What I have since learned is that, in spite of years of training and supervision, clinical work is challenging. It is a bumpier road than our minds want it be. Therapy work can be full of pain, uncertainty, not knowing, and insecurities, even when you are an "expert." Being an expert is not "the" safe place or the miraculous ingredient. Therapy is like life; it is a myriad of emotions and experiences. The essential point is willingness to embrace them all. And there are no difficult clients, there are only difficult clients for ourselves. Each experience in therapy, even with challenges, is an invitation to delve deeper into ourselves and in life.*

REFLECTIVE Practice 9.2

Spend some time considering whether you have experienced each of the challenges covered in this chapter in your therapy work. Explore your initial reactions and the urges that arise when:

- You are working harder than your client
- You have a "talker" in front of you
- Silence is called for or is heavily present

- Nothing you do is good enough
- You feel intimidated
- You are working with a great big giant mind
- You are confronting issues of right and wrong
- Your client is suicidal

Ask yourself, *How did I categorize this client? What kinds of thoughts and emotional reactions did I have toward the client? How was this client like me? How can ACT interpersonal and intrapersonal processes help in these situations?*

The Heart of Now and the Door to Next

The challenging behavior of clients may pull therapists toward heavier use of technique, particularly if the client brings multiple crises or difficult interpersonal issues to session. Therapists have varying degrees of skill in working with these challenges in therapy; however, none of this skill is exhibited in a vacuum. It occurs in a context of intrapersonal awareness, interpersonal relationship, and awareness of change over time. What sets a more skilled ACT therapist apart is the ability to rapidly conceptualize the function of the behavior in terms of the six core processes and respond to it within the interpersonal process, recognizing how the immediate work and consequences support a goal or goals for therapy. Whether it is a particular kind of behavior or some of the behaviors explored in this chapter, the invitation is always to move toward your client, having a heartfelt presence and working from a position of kindness and care. The ACT therapeutic stance—open, aware, and engaged—and your therapeutic presence will be important. It is our ability to embrace these challenges that provides a foundation of connection. Here, a strong therapeutic relationship can be established as two separate people merge into humanity. Let's turn to chapter 10 and explore the ACT therapeutic stance.

CHAPTER 10

The Stance of the ACT Therapist

I have just three things to teach: simplicity, patience, compassion.
These three are your greatest treasures.

—Lao Tzu

The stance of the ACT therapist, which I have touched on throughout the book, is likely the most important underlying quality that feeds ACT's heart. Indeed, in engaging your personal work with ACT and connecting to its processes as you compassionately and authentically work with your clients, holding the stance explored here, you will discover that *you* are a full participant in the heart of ACT—its wisdom and heart are reflected by and in you.

"Stance" can literally mean two things, and both are relevant. It can be defined as (1) an adopted position with respect to experiencing mental or emotional events, and (2) a position of the feet. It is a place occupied by experience and body, and it entails a particular kind of relationship with yourself and others. The "adopted" stance requires awareness and is intimately linked to the ACT Core Competencies for the Therapeutic Relationship (see Luoma et al., 2017). The position of the feet is explored through being bold (taking action, stepping forward even when difficult) and through speaking from the heart (connecting to self and other at the level of feeling with an authentic, simple, and sincere presence) by the therapist in therapy. In its fullness, the stance of the ACT therapist is not separable into categories. Rather, the stance is a set of interwoven concepts that are tied together—feet and heart—in a whole cloth by the threads of what it means to engage in a relationship designed to compassionately support another being in creating a meaningful life.

In considering where to fully discuss the stance of the ACT therapist in this book, I settled on placing it here, in the last chapter, with the hope of tapping into the *peak-end bias*. The peak-end bias is a psychological heuristic wherein people connect to and remember an experience based on its peak and its end (Fredrickson & Kahneman, 1993). The snapshots of these two experiences are carried forward as we leave an experience, in this case exploring the heart of ACT. As you move forward in your ongoing exploration of ACT, moving toward mastery and the heart of ACT, it is my hope that the ACT therapeutic stance will be, as an end experience, one that remains foundational and one that you remember.

Adopting a Position with Respect to Experiencing: Broadening Open, Aware, and Engaged

The ACT therapeutic stance can be cultivated and begins, as we explored in part 1, with personal practice. It entails bringing awareness and compassion to your own experience as well as stepping forward in your life, and holding that others can do the same. As an adopted position, it is embodied and is about *presence*. You are "communicating" that you are firmly here. The best metaphor I can think of to represent presence is that of a tree. Solid and firmly rooted in the ground, it does not lose its "treeness" despite the changing seasons and weather. It is here—present. This embodied stance, then, is about a full connection to self-as-context. You are whole, and nothing can be said or felt that strips that wholeness. You remain rooted.

This sense of presence is instantiated by selecting willingness (i.e., I choose to be willing) and engagement from the larger sense of "I." You can carry this presence across therapy in general and you can bring it to each moment. I am willing over time and I am willing here and now. This presence can be seen and felt in those here-and-now moments when a client expresses deep or difficult emotion and the therapist is able to be open and still, and to actively choose to experience what they feel in response, while remaining engaged. For instance, if a client is speaking of a quite difficult trauma—sharing a story that not only is challenging to hear but also evokes feelings of anger and disgust—the therapist's work is to sit in that moment: grounded firmly in compassion; holding that the client is much larger than the story; defusing from judgmental thoughts, theirs and the client's; remaining willing to experience what is there to be felt; and committing to stay present—rooted, standing solid in the experience.

Practicing Self-Awareness and Flexibility in Service of Other Awareness

Key to the ACT therapeutic stance is the ability to be self-aware. I have spent a fair bit of time exploring the relevance of practicing and growing personal self-awareness as part of mastering ACT. In addition to recognizing your experience as it unfolds and growing your capacity to observe thoughts, feelings, sensations, history—to the extent these can be remembered—and how they play out in your life today, cultivating self-awareness also means being present to other aspects of your experience, especially in your therapeutic presence. It means understanding your behavior, knowing when you are fused, avoidant, or out of touch with your values, and then shifting out of mind and back to values-based action. It is the ability to recognize when you are caught in old stories that hold you back or fears about what might be, and then to gently shift to "now." It is your capacity to acknowledge personal patterns of behavior and the contexts in which they emerge such that freedom to choose a behavior in any moment is available so you might act in other ways. But self-awareness also means being connected to and aware of how others perceive you and the impact of your behavior on others, in the moment, and across time. It means slowing down and showing up to your intrapersonal experience as noted, but it also means slowing down and showing up to your interpersonal experience—being conscious of you *and* other. There are layers of context to be aware of here, but you and *your impact on other* are part of this process. Therapy will elicit certain behaviors from the client, as your response will elicit behavior from them, and this cycle of interaction will be inside of a developing therapeutic relationship context that emerges over time. How are you perceived by other? Can your unwavering presence be experienced in the room? How does your presence influence this process? Let's explore further.

Practicing and cultivating self-awareness can assist you in discerning when to act—and not act—on an emotion, thought, or sensation, perhaps helping you to break longstanding patterns of behavior (by being able to "see" the patterns) that have caused difficulties in your life or that have been present in your therapeutic work. Practicing self-awareness can also assist you in discerning when to act on experiences in relation to how you are perceived by others and in recognizing your impact on other. I am emphasizing perception *by others* and *impact on other* quite intentionally here. In our personal awareness practice, we may turn to some of the more common efforts such as awareness to the breath or body. We may focus on the five senses or observe our own thinking or that we think. We are less often, it seems, mindfully aware of how we are perceived or of what our impact is on others. Self-awareness in the ACT therapeutic stance entails both. Exploring impact on other is informative to your stance.

One question you might ask in this process is "How do you know that you are not self-aware?" The first thing that comes to mind is that you wouldn't be reflecting on self-awareness, but there are other signs indicating lack of self-awareness and awareness of impact on other. These might include a number of things linked to a single ACT pathological process (e.g., avoidance, fusion, past and future focus, self-as-content) or larger patterns of behavior that entail multiple ACT processes such as being passive-aggressive, needing to be in control, being defensive, or being a bully. Impact on others in each of these scenarios may be minimally examined, if at all. Lack of self-awareness might also include arrogance, stubbornness, poor boundaries, or perhaps neuroticism. Each of these, too, captures both a lack of awareness of self and a lack of awareness of interpersonal process or impact on other. I would venture that none works well for holding an open and engaged stance in therapy and thus a sense of presence. I would also venture to say that some of these are experiences we all encounter from time to time, not just in others, but in ourselves. There may be other behaviors that would indicate a lack of self-awareness and awareness of impact on others. Whatever the case, our work on self-awareness is never-ending and includes paying attention to the experience of other, asking about how our behavior is perceived and received. This type of awareness will inform the ways in which we work on or cultivate our personal therapeutic stance—hopefully letting go of behaviors that pull us away from cultivating an open presence.

Engaging in some form of self-reflective practice about your interpersonal patterns, such as practicing different forms of perspective taking, may prove valuable in this endeavor. One form is simply to take a self-reflective perspective: you seeing you. This involves observing your responses—and their workability—in different interpersonal contexts (again noticing your own antecedents, behavior, and consequences) and being aware of your emotions, thoughts, behaviors, and so on. The perspective you gain will guide you in considering the kinds of choices you would like to make based on what you learn, both in your personal relationships and in therapy. A second form is from the perspective of others: you "stepping into" someone else seeing you. This still has the quality of you seeing you, but you have now layered in how a particular person might see or experience you. You might discover, for instance, that they seem more or less connected to you, or more or less forgiving of you, than you are of yourself. Or they may seem more or less impacted by a particular thought or emotion than you might find yourself impacted by. You may find that they trust you less than you trust yourself, and so on. This kind of information can be useful in guiding you in therapy and testing hypotheses about your impact on others.

Depending on what you discover, use of curiosity, slowing down, and choosing can all be part of behavior change if you find the impact is not what you intended.

REFLECTIVE Practice 10.1

Explore the following questions as you consider your impact on others:

- What role does my impact play in creating a holding space?
- Am I experienced as open?
- Is there anything I want to change?
- Is there anything that keeps me away from a "rooted" presence? And is it something I want to work on?

Practicing mindfulness (I/here/now awareness) and other forms of perspective taking will cultivate your capacity to detect experience and flexibly respond to clients from inside an open stance or rooted presence. Working to have an ongoing awareness of you and the client's experiential state will support the capacity for an ongoing effective responsiveness to the client's experience. As John Kabat-Zinn (1994) noted, "our habitual unawareness and automaticity are exceedingly tenacious" (p. 8) and thus the need and effort to practice should be exceedingly tenacious as well. Avoiding reactivity and defensiveness with our clients, knowing where our "issues" end and theirs begin, using our mind and our gut for understanding and sensitivity, can only be garnered through conscious awareness of self and other. This awareness work will assist in speaking from the heart (see below). Knowing what you feel, think, and sense—being aware of your experience and what elicits and influences your behavior as well as its impact on others—is all part of an authentic, simple, and sincere presence in therapy.

Manuela's Reflection

The ACT therapeutic stance not only has to do with what we do verbally and with our feet, but also seems, the way it is presented, like an atmosphere that we emanate or send forth. It seems that it is something that is more of a gut feeling, and it provokes a sense of "being held" by our clients. It is a subtle presence that needs to be addressed more broadly in ACT and worked on by ACT therapists.

What you are speaking to here is very important—our impact on others plays a significant role in therapy. We all have a learned interpersonal repertoire that will show up inside the therapeutic relationship. We tend to punctuate relationships with derived interpretations, and our interpersonal exchanges are predetermined in manner. So questions I usually ask myself are What am I bringing to this relationship? What ways might I interact with this client that are solely about me? How can I best serve my client? *It is here that when I am honest in my noticing, a choice about how to proceed emerges.*

Carlton's Reflection

I really liked this section on self-awareness and the therapeutic stance. It would be useful for me for this to be operationalized a bit more. What are the key questions I need to ask of myself about my impact on other in the context of the therapy room? I need this to be simplified to a couple of key prompts, as it is such a demanding thing to do in this context. Are there ways of going about doing this? Maybe slowing things down, asking permission of the client to pause so that I can access these types of awareness? Asking the client to help me to do this? Isn't switching between self- and other awareness within the therapy room incredibly demanding? Maybe this could be acknowledged?

Carlton's reflection, above, sets us up nicely for the following Reflective Practice, which will be part of the journey in answering these questions. But it is particularly important to note that it is truly a journey, a process. Being alert to self and other within the therapy room may initially seem demanding, but as you develop skill, it becomes less of a demand and more of a gentle practice of noticing. Ultimately, it is my hope that therapists get so practiced at noticing that they can do both fairly easily and at the same time. The work of recontacting self-as-context over and again is helpful here, and ongoing recognition of other as context will also be part of this process. Regular mindfulness and compassion work can only assist.

REFLECTIVE Practice 10.2

Start the journey of reflective practice by exploring with others, in supervision or with colleagues, the questions below. Be honest in your exploration and curious in your approach. I invite you to track your behavior at key moments in therapy (or even personal relationships) before reflecting on these questions. Note where you might want to make changes as you cultivate your therapeutic stance:

- How am I perceived by other? How do I think I am perceived by my clients?
- Can my unwavering presence be experienced in the room?
- How does my presence influence this process? Am I experienced as open and whole? (Ask others who know you and seek honest feedback.)
- What role does my impact play in creating a holding space for clients?
- What am I bringing to this relationship?
- What ways might I interact with a particular client that are solely about me?
- Do I *really* hear my clients?
- What do I feel, what do I sense, and what do I think when I am fully present to my client?
- What do I think they feel, sense, and think when they are really heard?

Based on what you learn in your reflection, is there anything you want to change? Is there anything that keeps you away from a "rooted" presence that you would like to shift? And is it something you want to work on in exploring how to best serve your client?

Developing the ACT Core Competencies

Developing your ability as an ACT therapist includes engaging the ACT processes through the lens of the therapeutic relationship and considering the associated core competencies in mastering the model. These core competencies are fully explored in *Learning ACT* (2nd ed.; Luoma et al., 2017), in which readers are given the opportunity to complete exercises to help them understand and personally connect to the skills noted. Given this, I will not review them all or in detail, but instead I will reflect on a few and their significance in the ACT therapeutic stance.

The figure that follows depicts a model of the ACT therapeutic relationship. The most important part of this figure isn't the two hexaflexes drawn at each end; it is the lines drawn in between. They represent fluidity and interconnectedness. Each of the six core processes is linked to the other six, and all are flowing between client and therapist. When this is working masterfully, you can observe an entire session that is ACT consistent yet hasn't a single exercise, designated protocol, or specified tool in it. Not to say that those couldn't be part of a masterful implementation of ACT, but again, fluidity isn't about exercises and tools; rather, it is about process, including the overarching and ongoing, interpersonal, and intrapersonal processes. Fully relevant to these processes are a few core competencies that speak directly to the therapeutic stance.

A Model of the ACT Therapeutic Relationship

Practitioner Psychology — Interpersonal — *Client Psychology*

Intrapersonal — Intrapersonal

Therapy Interactions

Time / Overarching and Ongoing Process

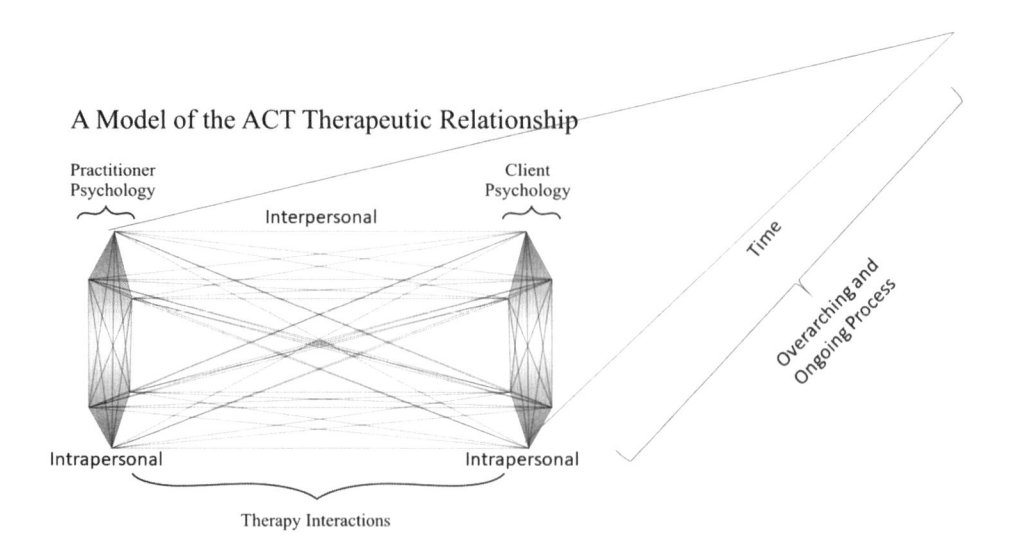

CORE COMPETENCY: THE THERAPIST SPEAKS

The ACT therapist speaks to the client from an equal, vulnerable, compassionate, genuine, and sharing point of view and respects the client's inherent ability to move from unworkable to workable responses (Luoma et al., 2017).

Equal. One of the most appealing aspects of ACT to many therapists and myself is its focus on humans as humans. We are all in this together and experience our own measure of suffering and joy. To work with the client from an equal position—both are human—is humbling by nature. Stepping into the therapy room as a person is different from stepping in as a doctor or other degreed person. You and the client are of the same value and it is your job to keep it balanced. Moving into a one-up position is about fusing with a concept of yourself. When this happens, flexibility decreases and the client is diminished. Equality is accessible from a willing, defused, self-as-context (perspective) position. Additionally, we all literally stand on the same ground and we will have the same end. Working from here not only removes the terrible beast of disordered and abnormal that often eats the client, it also grounds us in our own unassuming space, resulting in a respectful relationship with the client.

Vulnerable. Being in this space is also about being vulnerable. Being vulnerable, most simply stated, is about being open to pain. It literally means that one is susceptible to being wounded (dictionary.com, s.v., "vulnerable")—one can be hurt. It is the opposite of being defended and closed off. Being vulnerable with your client means modeling willingly holding pain, discomfort, anxiety, and all things difficult to experience. Any hurt that rises will also fall. You and your client are the context for experience. The self is not literally wounded; the self (as context) remains intact and is open to having the experience of pain, as an experience. This lends itself to solid presence—no pain will destroy the self (no season will rob the tree of its "treeness").

I have been asked if it is necessary for the therapist to be accepting and defused when doing ACT (or when simply living). It is vulnerability that leads me to the answer: yes. If we, ourselves, are willing to have pain, then we model and communicate the same to the client. Vulnerability is where intimacy is created and deep understanding and connection are born. If you are willing to experience in this way, then you can also "position the feet" to engage compassion.

Compassionate. Sitting with yourself or another while in deep pain and not shrinking away or pulling back is what it means to be compassionate. From a sense of self that is larger than any internal experience, compassion is possible. Pain from this perspective is experienced, its nuances and changes across time. Defused and open, compassion has a quality of kindness and giving. You are there and present for your client. Not only is compassion an important part of connection and healing in therapy, as commonly expected by therapists, but science has also noted its benefits. For instance, research has shown that compassion cultivation is associated with decreases in worry and emotional suppression while supporting adaptive and flexible functioning (Jazaieri, McGonigal, Jinpa, Doty, Gross, & Goldin, 2014). As well, compassion has been linked to better overall health (see Terry & Leary, 2011). Being compassionate doesn't mean that you can't be honest and direct, or that kindness is delivered at all costs; remember, compassion without wisdom is not compassion. Cultivating compassion for yourself and your therapeutic work is fully part of cultivating presence.

Genuine. To be genuine is to be free from pretense, affectation, or hypocrisy (dictionary.com, s.v. "genuine"). It means to be real. This is important on a couple of levels. First, it is directly linked to values such as authenticity and sincerity—it is about being true to yourself and the client. Of course, this doesn't mean that you share every thought and feeling with the client. But the ones that you do are true. They are accurate as to what you are experiencing. Second, being genuine is about being free from affectation. I have seen this only on a few occasions, but from time to time, therapists have misinterpreted compassion to mean that they need to be overly affected when listening to client's stories. Better to cultivate actual compassion than to mimic an idea of compassion. Third, and this is part of why I consider it important to be practicing ACT in your own life, being genuine means that you are free from hypocrisy. Inviting others to be open to their emotion, thought, and sensation experience, if you are not at least working at it, carries a kind of pretense. It is my experience that those therapists who are invested in excessive and misapplied control of their internal experience have difficulty implementing ACT, whether they are trying to master it or not.

Sharing. To share is to have or experience something together. In this sense, it is not only the client who reveals their emotion and thoughts, it is the therapist as well. The therapist reflects their internal experience in session and is aware and speaks to the shared joys and pains of humanity. The client is not alone. Together, client and therapist experience an interpersonal recognition of sameness—humanness.

Equal, compassionate, vulnerable, genuine, and sharing all feed a point of view—a stance. This stance is realized when we recognize the client's capacity to make change, holding them as capable and able to respond, respecting their inherent ability to move from unworkable to workable responses. Something can be done (some have heard me say, "If Stephen Hawking can use just his eye movements to communicate with the world, then something can be done"). Sometimes, even the smallest change matters.

CORE COMPETENCY: HOLDING THE TENSION AND BEING BOLD

ACT-relevant processes are recognized in the moment and where appropriate are directly supported in the context of the therapeutic relationship (Luoma et al., 2017).

Mastery of ACT involves identifying patterns, as well as moments in the session, of fusion, avoidance, being overly focused on the past or future, buying a concept of oneself, lack of or impulsive action, and values-inconsistent behavior tied to problems in living for the client. Once identified, the therapist intervenes using intrapersonal and interpersonal processes as well as multiple core processes, individually or simultaneously, depending on needs and timing, to assist the client in making positive life change. Sometimes identifying these patterns means becoming aware of tensions among different aspects of the therapeutic process. Being able to hold these tensions is found in the therapeutic stance.

Holding the tension. Skilled ACT therapists have the capacity to hold multiple tensions in the therapy room at the same time. This is also part of why awareness is particularly important for the therapist. The work here is about being able to discern when and how to act on events occurring within the process of change across time, interpersonally, and intrapersonally. When do you as a therapist intervene in the midst of an often rapidly moving therapeutic process? The different kinds of tensions that can be present in the room range in nature from maintaining silence (and for how long) versus speaking, to establishing more boundaries versus less, to interrupting and redirecting versus quietly listening, to reinforcing a behavior versus not, to focusing on what is being said versus what is not being said, to confronting versus not confronting, to self-disclosing versus not.

Considering these different kinds of tension and when and how to act on one versus the other requires grounding in theory, a good yet flexible case conceptualization, a functional understanding of behavior, knowledge of self, and a patient presence in the room. One of the ways in which I practice holding tension and either leaning in or leaning back, depending on what is called for, is simply by slowing down—and showing up to the experience and process in session. It is difficult to discern, even if you have a great case conceptualization, whether to confront a client or hold back if you are missing the emotional, psychological, and physiological cues between you and your client. Many therapists are able to clearly state what the function of a particular behavior seems to be. But knowing when and how to intervene is more challenging. For instance, I once sat in silence with a client for forty minutes, holding the tension between silence and speaking. Doing this might sound excruciating to most therapists. Trust me, it was no picnic. It was, however, just the thing needed to give the client space to generate behavior (i.e., anything that would create connection, asking, "How are you?" for instance). I held the stance of open and engaged, watching my own anxiety rise and fall and my mind blather on about what was happening. I simply held the tension, moment after moment. I slowed down so that I could "see" what was needed. I showed up to the experience in the room: mine, his, and ours. No need to run or rush forward.

Not all work that is about holding tensions will involve such radical events. However, it's important to practice awareness in the room so that you have the capacity to make a functional intervention that is relevant to the moment. When the therapist is rushing or feeling as if they need to make an intervention before the session is over, "missed opportunities" occur. Therapists tend to "run over the top" of opportunities to intervene by either verbally holding back (not speaking to what needs to be said), being too polite or a bit fearful about interrupting, or talking more than is needed. If you can practice patient awareness during session, lightly keeping the conceptualization and where you are heading in the background, holding the multiple tensions so as to not solve them too quickly, then you will more likely see when to intervene and which ACT-relevant move will best fit. You will be able to discern whether to lean in or lean back, in either case keeping a steady presence.

Being bold. Inviting clients to take bold moves has long been a part of ACT. One of the first trainings I did on being bold, titled "Putting Boldness into ACTion," was in 2004. What I noted at that time still stands: risk is part of change. The word "bold" can be defined as showing a readiness to take risks, a readiness to be daring or act with valor, a quality of taking action in the face of fear (or anxiety, sadness, and so on). Similarly, being "daring" can be defined as having enough courage to take an action; it is not about the feeling of courage, but rather about the will to do something hard (including taking emotional and psychological risks). I mention these definitions here as they hold that not only should we invite our clients to make bold moves, but as therapists, we are invited to make bold moves as well.

In this context, being bold is generally meant for the therapy room. It asks this question: are you as a therapist willing to take risks and do bold moves when interacting with your client? It may mean moving in session in the face of your fears and anxieties about what will happen. It might include being willing to experience fear of evaluation or of getting it wrong. It might mean saying the very difficult thing that you didn't want to say. It may mean introducing paradox or being willing to interrupt your client. It also means that you don't always hit the target. It is okay to make a mistake. What is bold for you?

Not taking risks or being bold in therapy is often tied to personal avoidance. We sometimes don't do things in therapy because we are avoiding our own fears. In being bold, though, a quality of confidence, as it is literally interpreted, is engaged. It ties back to the therapeutic stance. Confidence means with fidelity. Fidelity is synonymous with integrity or trueheartedness. This kind of boldness then is authentic or genuine; it is linked to an action that includes holding experience in a truehearted way. Being bold is also functionally practical and promotes growth and movement in the session. Whatever bold move you make, it should line up with your personal values—it is about being a therapist in your own truehearted way in the service of helping your client. It is also worth noting that bold moves must, of course, always be ethical. Being bold doesn't mean behaving badly or that it is okay to break ethical boundaries.

Some therapists fear that being bold—saying and doing things in therapy that contain risk (e.g., paradox, physically moving around in the therapy room, kind but blunt interaction)—is not being compassionate. I would argue that it is the direct opposite. As noted earlier, compassion without wisdom is not compassion. Although challenging, saying things that clients might find difficult to hear is part of what we are doing in therapy. Consequating or predicting consequences for behavior is part of that process. Being bold, then, can support the ACT work and promote discovery. Revealing and bold comments and dialogue are a part of that process, as illustrated in this example.

Client: (*somewhat demanding*) I thought things would be working better by now.

Therapist: (*curious*) Can you say what you mean by "better?"

Client: Well, you know, I thought I would be over this problem. I thought it would be easier after a while.

Therapist:	May I ask which problem you are referring to?
Client:	My anxiety. I am just sick of it. It ruins everything.
Therapist:	So, if I am understanding you correctly, it's not better now, because you are not over your anxiety?
Client:	Well, yeah. And I know what you are going to say…I need to feel it.
Therapist:	Actually, you don't need to feel it. [*starting a bold move*]
Client:	What do you mean? Okay, let's go, let's make this happen.
Therapist:	I mean you have been here before. And you can go back.
Client:	I don't understand.
Therapist:	You used to hang out in your room and drink alcohol so you didn't have to feel anxiety.
Client:	What? Are you saying I should go back to drinking?
Therapist:	(*genuinely*) Well, it is up to you. If you don't want to feel, then you can do what you have done.
Client:	But I don't want to drink and hide in my room.
Therapist:	I don't want that for you either.
Client:	Well, that is what you are telling me to do.
Therapist:	Hold up. I am not telling you to do that. You can also choose to be willing to feel anxiety.
Client:	I just don't want to.
Therapist:	Then don't. [*Bold move*]
Client:	Well…what would we do in therapy?
Therapist:	(*firm but kind*) Not sure you need therapy under this circumstance. You already know what to do if you don't want to feel anxiety.
Client:	Are you kicking me out of therapy?
Therapist:	Absolutely not. I just don't know how to remove your anxiety. I don't know how to make it so that you don't feel. And I don't want to pretend that I do.
Client:	I just hate feeling that way.
Therapist:	I know.

Sometimes therapists, both those new to ACT and seasoned therapists, say (in response to example dialogues like the one above), "I wouldn't have thought to say that," or "I wouldn't say that; it would be too scary." What if the client left therapy and went back to drinking? Or left therapy because of this conversation? Or what if having this kind of conversation would lead to too much anxiety or be too "mean"? It might make the client feel bad. Or it might be too hard for the client to hear. Needless to say, being bold is not free license to do anything in therapy. Being bold doesn't condescend, patronize, abuse, abandon, manipulate, or lie. Being bold in therapy *is* compassionate. It has integrity and authenticity. It is not about being mean, but it can certainly contain anxiety. Ultimately, the work of being bold is about being truthful with your client in ways that you may have been afraid to be. Not only does it model courage for the client, it gives permission to be bold back. Finally, it is creative and expansive. Making bold moves in therapy sets up rich and generative opportunities to work with clients in new and open ways and holds hope for the capacity for change.

REFLECTIVE Practice 10.3

Part 1. Think of a client you have been wanting to say something to that you have been avoiding saying—someone with whom you find yourself feeling anxiety, worry, or fear when you consider sharing something that would be truthful (given the context) but hard. Ask yourself these questions:

- What would it mean to be bold and speak from the heart in that moment?
- What would I be willing to feel if I were bold?
- How could I say this in a way that was honest but ACT consistent?

Consider taking action on this bold move. The best thing to do is slow down. Pause, breathe, connect, risk.

Part 2. Explore these questions concerning bold moves:

- What would be a bold move (or moves) for me in my therapeutic work?
- What would be the barrier to making that move (or those moves)?
- How do I think I would change as a therapist if I made that move (or those moves)?
- How could I be bold today? What would my generative, creative side like to do?
- Who will I report to about my bold move?

Be bold in one aspect of your life as a therapist with a time commitment to get it done. As the old ACT metaphor goes, jump off a piece of paper or a bungee tower, but do jump.

Positioning the Feet: Choice and Existentialism

As feeling, sensing, acting beings, as conscious beings untethered from our concepts of ourselves, we are free, but also not. Despite our capacity to choose, we are bound by certain limitations in life. There are current environmental and historical limits; there are limits of where and into what context we are born. There are limits of nature. But the most glaring limitation of all is death. We cannot choose to live forever. This knowledge is perhaps a great burden, and many will suffer with this knowledge, spending days and years fretting about its impending nature or avoiding the reality. This knowledge, however, could also be a great blessing—it is possibly the only knowledge that leverages how we will choose to be during our short time alive.

Bringing Existential Angst into the Therapy

Existential angst, or fear of one's own freedom and responsibility, challenges both client and therapist. In its simplest form, the knowledge that we can create our lives (within the limitations noted above) is both wonderful and scary. It means that we can make a different choice at any point in time as we move forward in our lives: a choice that can fully alter our life. It thoroughly removes the oft uttered "I can't" and replaces it with "I choose not to." Choosing not to is the full responsibility of the individual. Again, this is beautiful and frightening.

Bring the notion of full responsibility for your life to mind: You are responsible for whatever happens in your life. Notice what happens here. Take time to truly and honestly connect to what this means. Be aware of each of the possible places where things could be different if you were to make a different choice. Consider all of the possible places where a slight, yet chosen, turn in your path could mean something very different for you and how you live now. This can be scary in the sense that it means you can choose, right now, to do something completely different with your life. Imagine this scenario: you set this book down, and without putting on a shoe, you walk to the front door, open it, and walk away, never to return to this life you have now. That is possible. And scary. And a part of the ultimate absurdity that many existentialists have written about—the meaningless of it all. See if you can contact the anxiety that arises with this knowledge. I invite you now to also contact something else: this kind of freedom and responsibility also means that you can create.

You and your clients have a very short time to be alive. And the fact is, you are here. And you are alive. So rather than wrestle with the why of this fact, I suggest creating the how and what of it. Stated differently, what meaning will you create inside of this great responsibility and freedom? I bring this knowledge, this existential angst, this freedom and responsibility into the therapeutic stance. This means from the most human and vulnerable place, I want my clients to live. But not just any life, rather, a life that is fully linked to what is meaningful to them, a life fully encapsulated by what they care about and want to create. I am invested and persistent in supporting them in taking action and responsibility in their lives, in acting on the freedom that is available to them. I hold this as part of the therapeutic stance, not just because ACT has values and committed action—I hold this as part of the stance because we die.

And we are all running out of time.

Listening with Awareness, Responding from the Heart

Notice the difference between the two ways of listening and responding in the following two examples.

Time 1: Listening Well and Responding (therapist notices all things in parentheses)

Client: (*soft voice, stammering a little*) It is…it is hard for me to talk about my childhood. I…I…I really hated being a kid. I got my breasts early…I was about twelve (*long pause, looks up*). My dad used to force me…to stand in front of the mirror (*seemingly holding back tears, takes audible breath*)…naked. He would force my shoulders forward and back, showing me my posture and what it did to my breasts (*pause*). He would cup my breasts with his hands…(*looks down and starts to cry*).

Therapist: (*after a short pause, speaking softly*) That sounds…really terrible. I am so sorry that happened to you. Your dad was wrong for doing that. How old did you say you were?

Time 2: Listening with Awareness and Responding from the Heart (therapist notices all things in parentheses)

Client: (*sitting slightly hunched over and forward, soft voice, stammering a little*) It is…it is hard for me to talk about my childhood. I…I…I really hated being a kid (*quickly looks down and then up again*). I got my breasts early…I was about twelve (*long pause, looks down and appears to be biting inside of lower lip, shifts in chair, clears throat, looks up*). My dad used to force me…to stand in front of the mirror (*voice is halting, seemingly holding back tears, takes audible breath*)…naked (*looks down, looks up, appears to be biting inside of lower lip, next thing is said quickly*). He would force my shoulders forward and back, showing me my posture and what it did to my breasts (*pause, sighs, looks down*). He would cup my breasts with his hands…(*looks down and starts to cry, hunches over slightly, further drawing her chest in*).

Therapist: (*after a longish pause, sitting with client while she cries, without interrupting or moving to comfort, then, being aware of her own experience, speaking softly*) I feel my heart sinking heavily into my stomach; I feel a kind of sadness and shame all at once when I hear about what happened to you (*followed by a long silent pause*).

The responses made by the therapist in time 1 and time 2 are both thoughtful responses. However, with time 2, the therapist is paying closer attention and is noticing the body language of shame and sadness; she is willing to feel and speak to her intrapersonal experience, sharing it with the client. In the second version the therapist is more attuned, more connected and empathic, while the first seems a bit more "clinical." The second is what I refer to as listening with awareness and heart. The therapist is noticing multiple experiences and listening intently to verbal behavior that is spoken using both language and body. Here is where listening with awareness and responding from the heart can begin.

I started doing an exercise in ACT workshops a few years back that has become a personal favorite. Participants are paired up, and each takes a turn playing different roles: therapist and client. The "client" is asked to work on a personal experience of pain. I ask the "therapist" to listen quietly for a few minutes while their partner talks about their struggle, and then I invite the therapist to "go to work," to do what they might typically do to help this person as they engage in a conversation about the issue presented. The therapist is given about two to two and a half minutes to interact with the client. In round two, I again ask the therapist to listen quietly, but this time with full awareness, with mindfulness to the moment, and when cued to respond, to do so *from the heart*. This means the work to do this time is to say to their partner what it *felt* like, for them personally, to hear this story of pain: to respond from the heart, not the head.

When speaking from the heart the words are fewer, are more metaphorical, and take less time. The words are about what is truly felt when you hear another's pain. Almost without fail, following the exercise, participants report that the "client" felt more connected, better understood, listened to, and empathized with in time 2. The therapist's response in time 2 isn't fancy, it doesn't contain interpretations or questions, it doesn't call for exercises or a focus that is designed to get somewhere—rather, it is simple, it is open, it is authentic. More tears flow here by far than in time 1.

To be fair, most people report that time 1 was helpful as well. However, the work there tends to lean more toward problem solving, reflective listening, or further questioning about the struggle at hand, all of which might be fine work to do with a client. But it also turns out that simply telling the client how it *felt* (from their hearts) to hear about what the client experienced has an impact too. A powerful one. But the "hearing" also has to be powerful, the awareness broad, taking in both client and self.

You don't have to share your every internal experience about your client's story or respond from the heart multiple times a session, or even in every session. Rather, doing this when you see and hear the client's pain is part of what it means to empathize and communicate authentically, one human being to another. Bringing this kind of presence to the therapy, I would argue, is part of the ACT stance. It keeps the therapist intimately connected to process and experience.

This kind of interpersonal process can truly be done only when the therapist listens to the client with full awareness. I often describe this type of listening through metaphor. It is a listening that is done "with every bone in the body." It is a focused attention on the client's whole being—their body language, their words, the sound in their voice, their posture, and their "presence." It is about being able to detect the slightest quiver or catch in the voice, a small change in facial expression or a slight move of the shoulder. "Seeing" all of these are part of listening with every bone. Each communicates something and tells you about the client's current state and need.

Inside of that listening, as noted, you also carry a smaller awareness of yourself. Maybe 25 percent of your attention is on your own experience. Noticing what is happening for yourself when you encounter all that is happening in the person sitting across from you allows you to speak from the heart. It also tells you something about how you might respond and about what is needed. It brings your and your client's inner worlds closer together. It's not foolproof, but it is authentic and experiential; it is out of the head and into the interpersonal life that is occurring between you and your client.

Deepening the ACT Therapeutic Stance

Taken together, self- and existential awareness, the ACT Core Competencies for the Therapeutic Relationship (Luoma et al., 2017), holding the tension, being bold, listening with awareness, and responding from the heart define and broaden the ACT therapeutic stance. They expand the stance by pointing to a kind of simplicity that is found in and linked broadly to the authenticity inside of consciousness and knowing that we will die, allowing the therapist to be genuine, vulnerable, and sharing—boldly and from the heart.

The Heart of Now and the Door to Next

In working on your personal therapeutic stance, and from an ACT perspective, I invite you to expand your presence in therapy, by stretching into and cultivating patience and awareness. With this practice you will better identify the tensions that need to be held, considered, and acted upon or not. You will better connect to the ongoing flow of process and experience. You will better communicate in an empathic and compassionate fashion. You will better recognize freedom and the pull to live as we move toward our own death. You will better bring heart and wisdom to your ACT work. In so doing, you can fully listen and assist, using processes that promote openness, awareness, and engagement, supporting others in choosing to create meaningful lives—head, heart, and hands.

EPILOGUE

Owning Your Life; Living with Heart

There is no cure for birth and death save to enjoy the interval.

—George Santayana

Attention in therapy takes work. Weighing each word and understanding each sentence a client has to offer is an effort. Sitting with an individual's pain in the service of helping them to experience something larger, contacting a sense of self that is neither defined nor limited by their pain, involves a recognition, an awareness of what it means to connect with purpose in life, bringing it into the here and now. Awareness of our own death seems a necessary part of this work—not a morbid awareness, but a curious one. One that makes the reality of itself available, one that threads through the fabric of life—one in which death is fully a part of our existence. This recognition isn't easy. At times it is painful, and even ironic. I am reminded of a quote attributed to Robert Frost: "Forgive, O Lord my little jokes on Thee/and I will forgive Thy great big one on me." We live, we die. What we do with that little space in between matters. And what matters is there for us to create. What matters can be bound to rigid attempts to not feel pain; it can be bound to attempts at control of our internal life. Or it can be fluid, flexible, free—present to and savoring of each moment.

I recognize the challenge that lies within this venture. Attention in life takes work. But this is meant to be a gentle and steady process, returning as often as possible to this moment. Cultivating a practice that grows flexibility in the service of engagement is lifelong. There is no arrival. It involves assisting clients in a process of owning what they create with their feet. This is easier done from a place of wholeness—clients are whole as they are (as are you). They do not need to fix their emotions or thoughts before they choose their next step. Getting to this place with clients using ACT will include bringing technique and procedures to bear in the therapy room. However, these are not done in a vacuum. An ongoing process that involves a conceptualization of the client's avoidance, fusion, fears, and constrictions and the movement to flexibility will be part of this endeavor—and tucked inside of this arc of therapy is *you* and your relationship with the client. The intra- and interpersonal are woven together, balancing client, therapist, and intervention in a fluid and compassionate implementation of ACT.

I remember the precise moment I connected to a broader sense of self in a life-changing experience that brought both clarity and the possibility of choice and freedom. It was in this moment that I fully recognized the power of compassion and the fundamental movement of life. Taking a compassionate perspective toward my own pain, choosing kindness over criticism, brought a deep level of understanding to human experience, my own and others', and freed me to move in ways that I had not contacted before. As important was the awareness of movement, of the ongoing flow of life experience. It does not hold still. Every ounce of being, everything thought, everything felt, everything sensed, rises and falls. Life is a process. I rise, I live, I die. Following this moment of awareness, I asked, *If this is the course (and it is), then what is my meaning? How will I live?* Here lies the intersection of this powerful moment years ago and my forward path—my bumpy, windy, at times chaotic, joyful, painful path.

This moment occurred for me during an ACT exercise (visiting yourself as a child from the perspective of an adult, buoyed by self-as-context) at a workshop I attended years ago. In this experience, I "met" both a stable and a fluid sense of self. Both consciousness and myself as a process were present. Aware of joy and pain, I stretched across each and was neither. The quote "Be still like a mountain and flow like a great river," attributed to Lao Tzu, captured what happened. This powerful shift in relationship to my thoughts and emotions occurred; and the most critical process, *for me*, in ACT came to life: self-as-context. I am the context for experience; I am the experiencer.

The "child me" that I visited in this exercise was innocent enough, but already familiar with significant pain. The pain of life at that time was filled with authority, control, and bouts of violence. In visiting her from the perspective of "now" (seeing her in my imagination during the exercise), I recognized first that I was "seeing" her—that I was here, now, seeing her there, then. There was a larger "I" witnessing and extending back in time, shifting between both seeing and "being" this child. Second, I could also "see" that she needed something. She needed love, she needed acceptance. It wasn't available there, back then. But…I could make it available here, now. I could choose compassion in that very moment, standing in awareness and offering something. I chose to give. I met this young girl in a moment of clarity and offered her what she had not been given as a child. I made room for her with me. I would no longer fear her vulnerability. There was heart and wisdom in that moment. The experiencer is untouched by history, and what we do with our feet from here forward matters. In living this short life, how will I be with myself and others? What will I choose? It is in this place that even deep pain can be transformed. Not by changing the past, or feeling good, but by opening to the flow of life, by seeing the richness in beauty and failure, loss and love.

This recognition—that we are not our experiences, that we are the experiencer and that where we choose to place our feet creates meaning—can be brought to life in the work that we do. It can be embodied in a stance that is steadfast and consistent with a recognition that we are not our words. That we are more extensive than criticism and pain. Indeed, "the pleasure of criticizing takes away from us the pleasure of being moved by some very fine things" (attributed to Jean de La Bruyère). Living inside of the mind takes away from the moment, robbing us of the fine features in a flower, the wrinkled eyes of a smile, the wag of a dog's tail. Come back to I/here/now. Touch these qualities in life that make life itself.

And also engage the mind, in its ability to provide us with direction, in its ability to clarify and understand choice and values. Next, what I choose as I move forward will bring meaning to life. I believe that true wisdom lies here: being open to the moment, available to possibility and choice, aware of mind and unattached from its judgment, stepping into the fluid movement of a life. The invitation is always there—to step into your life. My hope is that in any pursuit of mastering and moving into its heart, in embodying the work done in ACT processes, you will find a fluid way to engage the six core processes, interpersonal and intrapersonal experiencing over time, in such a way that you will not only benefit your clients, but engage a personal path—a bumpy, windy, at times chaotic, joyful, painful path. Life with heart—and from the feet up.

Contributors' Notes

I would like to once again give a special thanks to Carlton and Manuela for their contributions as well as the time they gave in working on *The Heart of ACT*. Their willingness and patience were meaningful. In their final reflections, they note their experience—their journey. With gratitude, I welcome these reflections.

Carlton's Final Reflection

The idea that engaging with the head "takes away from us" a fuller engagement with the heart is probably, at least by now, nothing new. I personally have experienced the process of contributing to this book, primarily in the form of reviewing what Robyn has written, as a series of pulls on my head and my heart. An important word here is "series." These pulls have often come one after another—at times within individual sentences or paragraphs, at other times across pages and chapters. But at any single moment in time, I have usually felt a pull toward either what I think or what I feel.

More often than not, when the pull I have experienced engages my intellect, the result is a position of difference. I have tried to make any criticism constructive, and as I look back I see that this criticism has usually been aimed at specific aspects of ACT, the theory and practice, rather than anything unique to Robyn's position. It is probably this that has defined my very modest contribution to this book. For me, contributing has mainly been a process of exploring those areas of ACT that I struggle with at an intellectual level. It would be easy for me to push ACT further away based on these criticisms, were it not for the fact that there were definite pulls on my heart as well. Sometimes reading drafts of individual chapters also had a distinctly visceral impact on me. There were moments when I felt a real affinity with some of the things written, and I had an appreciation of an ethos, perhaps broader than ACT, which I really connected to.

Maybe at times, I have had some pleasure in stating where I have definite problems with ACT, particularly concerning the way absolutes are created, disseminated, and potentially conveyed to clients. Maybe it's also true that this has taken away some of the "pleasure of being moved by some very fine things." However, my contributions are an honest reflection of where I find myself at this particular moment in time. I just hope what I have written has helped, rather than hindered, the reader.

Finally, I do not feel qualified to say too much about what I personally think contributes to the mastery or fluid implementation of ACT, although I would make just one observation which is personal to Robyn, and which I think may be significant. It seems to me that to gain mastery of anything, one must be really committed to it. By "commitment" I do not mean a rigid adherence to something, where anything that is challenging is immediately dismissed without first being carefully considered. I am not talking about blind faith. Rather, I mean the kind of commitment that comes from having a very deep connection with something, almost certainly with head and heart. I wonder whether at least part of what leads to the elusive quality of "fluidity" we are constantly hoping to achieve as ACT therapists is actually this phenomenon of "commitment." I anticipate it is manifest in those micromoments of responding to our clients, it informs effective decision making as to where best to go next, and it conveys to the client a sense that the therapist has what it will take to help them. In summary, I am left wondering whether the therapist's level of commitment to ACT is ultimately a critical factor in their potential mastery of it. As in these words, attributed to Albert Einstein, "Only one who devotes himself to a cause with his whole strength and soul can be a true master. For this reason, mastery demands all of a person."

Manuela's Final Reflection

We are arriving at the end of this journey together and, still…there is no arrival. From here forward you will be adventuring with other companions in this vibrant and meaningful journey that is learning ACT. For me, to be part of the creation of this book has been an exquisite experience. I learned far more about ACT than I expected and found that this experience provided the opportunity to explore my understanding and connection to ACT personally and more deeply. I saw many different subtleties that make fluency and mastery of ACT possible. And the important word here is possible. *The embodiment of ACT in therapy and your personal life, if you choose, is viable. The work described in this book is part of that process. Is mastery possible? I think this book works its way through that possibility, layer by layer, and I hope that you will find it useful in this way too. Finding a way to distill the ACT stance, a form of being and doing that is embodied by the heart and wisdom, is my main quest and was the Sherpa that guided me and my contributions to this book. I hope, in your journey through this book, that you connected to a felt sense of how it is to master and fluidly implement ACT and more closely operationalize the stance and process of the intervention. It has helped me to grow as an ACT therapist, and I have the same hope for you.*

The process of contributing to this book was a communal adventure as well. I teach ACT a fair amount in Argentina, and with the arrival of every new chapter, I worked on an experiential translation, a way to bring what was written to life in training to support my students in their process of learning ACT. Some of my comments and contributions are infused with my students' voices and experiences. I was able to see how it helped them to grow, not only in ACT understanding and competencies, but also in the embodiment of ACT. Engaging with my students also reminded me that moving from learning to mastery is a relational process and best done with others. Learning through working together was a part of this book—Carlton, Robyn, and I each benefited from the others' insights and questions. The journey through this book included three different voices, three different perspectives, making the journey more exciting, engaging, and fun. The ongoing conversation among the three of us challenged me, and at times brought discomfort, but as noted, discomfort precedes growth. I gained from entertaining different points of view. I invite you to consider this kind of "walking together," studying and integrating the ideas shared in this book personally, but with others as well. And, from multiple views and diverse backgrounds, a genuinely contextual perspective, begin to embrace the journey to mastery. I hope you find that a deeper dive into the ACT stance and process will transform your therapy and your life.

References

Aarons, G. A. (2004). Mental health provider attitudes toward adoption of evidence-based practice: The Evidence-Based Practice Attitude Scale (EBPAS). *Mental Health Services Research, 6*(2), 61–74.

Ambady, N., & Rosenthal, R. (1992). Thin slices of expressive behavior as predictors of interpersonal consequences: A meta-analysis. *Psychological Bulletin, 111*(2), 256–274.

Ardelt, M. (2004). Wisdom as expert knowledge system: A critical review of a contemporary operationalization of an ancient concept. *Human Development, 47*(5), 257–285.

Baer, R. A., Smith, G. T., & Allen, K. B. (2004). Assessment of mindfulness by self-report: The Kentucky Inventory of Mindfulness Skills. *Assessment, 11*(3), 191–206.

Baer, R. A., Smith, G. T., Lykins, E., Button, D., Krietemeyer, J., Sauer, S., . . . Williams, J. M. G. (2008). Construct validity of the Five Facet Mindfulness Questionnaire in meditating and nonmeditating samples. *Assessment, 15*(3), 329–342.

Banks, A. (2016). *Wired to connect: The surprising link between brain science and strong, healthy relationships.* New York, NY: Penguin.

Bensing, J. M., Kerssens, J. J., & van der Pasch, M. (1995). Patient-directed gaze as a tool for discovering and handling psychosocial problems in general practice. *Journal of Nonverbal Behavior, 19*(4), 223–242.

Bradberry, T. (2015, August 25). 15 critical habits of mentally strong people. *Forbes.* Retrieved from https://www.forbes.com/sites/travisbradberry/2015/08/25/15-critical-habits-of-mentally-strong-people/#9355bcd717bb.

Cahn, B. R., & Polich J. (2006). Meditation states and traits: EEG, ERP, and neuroimaging studies. *Psychology Bulletin, 132*(2): 180–211.

Chödrön, P. (1991). *The wisdom of no escape and the path of loving-kindness.* Boulder, CO: Shambhala Publications.

Chödrön, P. (2000). *When things fall apart: Heart advice for difficult times.* Boulder, CO: Shambhala Publications.

Cohen-Cole, S. A. (1991). *The medical interview: The three-function approach.* St. Louis, MO: Mosby-Year Book.

Cox, D. (2015, April 16). Is your voice trustworthy, engaging or soothing to strangers? *The Guardian.* Retrieved from https://www.theguardian.com/science/blog/2015/apr/16/is-your-voice-trustworthy-engaging-or-soothing-to-strangers.

Eugenides, J. (2002). *Middlesex.* New York, NY: Picador, Farrar, Straus and Giroux.

Fatter, D. M., & Hayes, J. A. (2013). What facilitates countertransference management? The roles of therapist meditation, mindfulness, and self-differentiation. *Psychotherapy Research, 23*(5), 502–513.

Fauth, J., & Nutt Williams, E. N. (2005). The in-session self-awareness of therapist-trainees: Hindering or helpful? *Journal of Counseling Psychology, 52*(3), 443–447.

Frankl, V. (2006). *Man's search for meaning.* Boston, MA: Beacon Press. (Original work published 1959)

Fredrickson, B. L., & Kahneman, D. (1993). Duration neglect in retrospective evaluations of affective episodes. *Journal of Personality and Social Psychology, 65*(1), 45–55.

Geller, S. M., & Porges, S. W. (2014). Therapeutic presence: Neurophysiological mechanisms mediating feeling safe in therapeutic relationships. *Journal of Psychotherapy Integration, 24*(3), 178–192.

Gifford, E. V., & Hayes, S. C. (1999). Functional contextualism: A pragmatic philosophy for behavioral science. In W. O'Donohue & R. Kitchener (Eds.), *Handbook of behaviorism* (pp. 285–327). San Diego, CA: Academic Press.

Greenberg, L. S. (2015). *Emotion focused therapy: Coaching clients to work through their feelings.* Washington, DC: American Psychological Association.

Hall, J. A., & Bernieri, F. J. (Eds.). (2001). *Interpersonal sensitivity: Theory and measurement.* Hove, UK: Psychology Press.

Harris, R. (2019). *ACT made simple: An easy-to-read primer on acceptance and commitment therapy* (2nd ed.). Oakland, CA: New Harbinger Publications, Inc.

Hayes, J. A., Gelso, C. J., & Hummel, A. M. (2011). Managing countertransference. *Psychotherapy, 48*(1), 88–97.

Hayes, S. C. (1994). Content, context and the types of psychological acceptance. In S. C. Hayes, N. S. Jacobson, V. M. Follette, & M. J. Dougher (Eds.), *Acceptance and change: Content and context in psychotherapy,* (pp. 13–32). Reno, NV: Context Press.

Hayes, S. C. (2004). Acceptance and commitment therapy, relational frame theory, and the third wave of behavioral and cognitive therapies. *Behavior therapy, 35*(4), 639–665.

Hayes, S. C., Hayes, L. J., & Reese, H. W. (1988). Finding the philosophical core: A review of Stephen C. Pepper's *World Hypotheses: A Study in Evidence. Journal of the Experimental Analysis of Behavior, 50*(1), 97–111.

Hayes, S. C., Barnes-Holmes, D., & Roche, B. (2001). *Relational frame theory: A post-Skinnerian account of human language and cognition.* New York, NY: Plenum Publishers.

Hayes, S. C., Strosahl, K. D., & Wilson, K. G. (2012). *Acceptance and commitment therapy: The process and practice of mindful change* (2nd ed.). Oakland, CA: New Harbinger Publications, Inc.

Hayes, S. C., & Hofmann, S. G. (Eds.). (2018). *Process-based CBT: The science and core clinical competencies of cognitive behavioral therapy.* Oakland, CA: New Harbinger Publications.

Hendricks, M. (2009). Experiencing level: An instance of developing a variable from a first person process so it can be reliably measured and taught. *Journal of Consciousness Studies, 16*(10–12), 129–155.

Jacques, B. (2011). *The taggerung* (Redwall, Book 14). Random House.

Jazaieri, H., McGonigal, K., Jinpa, T., Doty, J. R., Gross, J. J., & Goldin, P. R. (2014). A randomized controlled trial of compassion cultivation training: Effects on mindfulness, affect, and emotion regulation. *Motivation and Emotion, 38*(1), 23–35.

Kabat-Zinn, J. (1994). *Wherever you go, there you are: Mindfulness meditation in everyday life.* New York, NY: Hyperion.

Kazantzis, N., Deane, F. P., & Ronan, K. R. (2000). Homework assignments in cognitive and behavioral therapy: A meta-analysis. *Clinical Psychology: Science and Practice, 7*(2), 189–202.

Kazantzis, N., Whittington, C., & Dattilio, F. (2010). Meta-analysis of homework effects in cognitive and behavioral therapy: A replication and extension. *Clinical Psychology: Science and Practice, 17,* 144–156.

Kornfield, J. (2001). *After the ecstasy, the laundry: How the heart grows wise on the spiritual path.* New York, NY: Bantam Books.

Kottler, J. A., & Carlson, J. (2014). *On being a master Therapist: Practicing what you preach*. Hoboken, NJ: John Wiley & Sons, Inc.

Lakoff, G., & Johnson, M. (1999). *Philosophy in the flesh: The embodied mind and its challenge to western thought* (Vol. 28). New York, NY: Basic Books.

Lakoff, G., & Johnson, M. (2008). *Metaphors we live by*. University of Chicago Press.

Lappalainen, R., Lehtonen, T., Skarp, E., Taubert, E., Ojanen, M., & Hayes, S. C. (2007). The impact of CBT and ACT models using psychology trainee therapists: A preliminary controlled effectiveness trial. *Behavior Modification, 31*, 488–511.

Luoma, J. B., Hayes, S. C., & Walser, R. D. (2017). *Learning ACT: An acceptance and commitment therapy skills training manual for therapists* (2nd ed.). Oakland, CA: New Harbinger Publications, Inc.

Mannion, P. (2009, July 20). How a felt-tipped pen saved the Apollo 11 mission. *EE Times*. Retrieved from https://www.eetimes.com/author.asp?section_id=4&doc_id=1283891.

Margison, F. (2001). Practice-based evidence in psychotherapy. In C. Mace, S. Moorey, & B. Roberts (Eds.), *Evidence in the psychological therapies: A critical guide for practitioners* (pp. 174–198). New York, NY: Brunner-Routledge.

Matthews G., Zeidner M., & Roberts R. D. (2004). *Emotional intelligence: Science & myth*. Cambridge, MA: MIT Press.

Mayer, J. D., Salovey, P., Caruso, D. R., & Sitarenios, G. (2003). Measuring emotional intelligence with the MSCEIT V2.0. *Emotion, 3*(1), 97–105.

McHugh, L., & Stuart, I. (2012). *The self and perspective taking: Contributions and applications from modern behavioral science*. Oakland, CA: New Harbinger Publications, Inc.

Mehrabian, A. (1972). *Nonverbal Communication*. New Brunswick, NJ: Aldine Transaction.

Mehrabian, A. (2009). "Silent messages" – a wealth of information about nonverbal communication (body language). In A. Mehrabian, *Personality & emotion tests & software: Psychological books & articles of popular interest*. Los Angeles, CA: self-published.

Mehrabian, A., & Wiener, M. (1967). Decoding of inconsistent communications. *Journal of Personality and Social Psychology, 6*(1), 109–114.

Nhat Hanh, T. (1976). *Peace is every step: The path of mindfulness in everyday life*. New York, NY: Bantam Books.

Nutt Williams, D. (2008). A psychotherapy researcher's perspective on therapist self-awareness and self-focused attention after a decade of research. *Psychotherapy Research, 18*(2), 139–146.

Nutt Williams, E., & Fauth, J. (2005). A psychotherapy process study of therapist in session self-awareness. *Psychotherapy Research, 15*(4), 374–381.

Pakenham, K. I. (2015). Effects of acceptance and commitment therapy (ACT) training on clinical psychology trainee stress, therapist skills and attributes, and ACT processes. *Clinical Psychology & Psychotherapy, 22*(6), 647-655.

Ramnero, J., & Törneke, N. (2008). *The ABCs of human behavior: Behavioral principles of the practicing clinician*. Oakland, CA: New Harbinger Publications, Inc.

Rogers, C. (n.d.) quote: Retrieved from https://www.goodreads.com/quotes/1171586-when-i-can-relax-and-be-close-to-the-transcendental.

Ryan, A., Safran, J. D., Doran, J. M., & Muran, J. C. (2012). Therapist mindfulness, alliance and treatment outcome. *Psychotherapy Research, 22*(3), 289–297.

Santayana, G. (1923). *Soliloquies in England and later soliloquies*. New York, NY: Charles Scribner's Sons.

Schirmer, A., Feng, Y., Sen, A., & Penney, T. B. (2019). Angry, old, male—and trustworthy? How expressive and person voice characteristics shape listener trust. *PLoS ONE, 14*(1), e0210555.

Strosahl, K. D., Hayes, S. C., Bergan, J., & Romano, P. (1998). Assessing the field effectiveness of Acceptance and Commitment Therapy: An example of the manipulated training research method. *Behavior Therapy, 29*(1), 35–63.

Tannen, D., Schiffrin, D., & Hamilton, H. E., eds. (2015). *The handbook of discourse analysis.* Malden, MA: John Wiley & Sons.

Terry, M. L., & Leary, M. R. (2011). Self-compassion, self-regulation, and health. *Self and Identity, 10*(3), 352–362.

Thompson, J., (2011). Is nonverbal communication a numbers game? *PsychologyToday.com*. Retrieved from https://www.psychologytoday.com/blog/beyond-words/201109/is-nonverbal-communication-numbers-game.

Tirch, D., Schoendorff, B., & Silberstein, L. R. (2014). *The ACT practitioner's guide to the science of compassion: Tools for fostering psychological flexibility.* New Harbinger Publications.

Törneke, N. (2010). *Learning RFT: An introduction to relational frame theory and its clinical application.* Oakland, CA: New Harbinger Publications, Inc.

Treasure, J. (2017, December). *Julian Treasure: 5 ways to listen better* [video file]. Retrieved from https://www.ted.com/talks/julian_treasure_5_ways_to_listen_better.

Villatte, M., Villatte, J. L., & Hayes, S. C., (2015). *Mastering the clinical conversation: Language as intervention.* New York, NY: The Guilford Press.

Vrij, A., Edward, K., Roberts, K. P., & Bull, R. (2000). Detecting deceit via analysis of verbal and nonverbal behavior. *Journal of Nonverbal Behavior, 24*(4), 239–263.

Walser, R. D., & Westrup, D. (2007). *Acceptance and commitment therapy for the treatment of post-traumatic stress disorder and trauma related problems: A practitioner's guide to using mindfulness and acceptance strategies.* Oakland, CA: New Harbinger Publications, Inc.

Walser, R. D., & Westrup, D. (2009). *The mindful couple: How acceptance and mindfulness can lead you to the love you want.* Oakland, CA: New Harbinger Publications.

Walser, R. D., Karlin, B. E., Trockel, M., Mazina, B. & Taylor, C. B. (2013). Training in and implementation of Acceptance and Commitment Therapy for depression in the Veterans Health Administration: Therapist and patient outcomes. *Behaviour Research and Therapy, 51*(9), 555–563.

Walser, R., & McGee-Vincent, P. (in press). How to develop your abilities as an ACT therapist. In P. Lucena-Santos, S.A. Carvalho, J. Pinto-Gouveia, M.S. Oliveira, & J. Pistorello (Eds.). *Manual prático internacional de Terapia de Aceitação e Compromisso* [International Practical Handbook of Acceptance and Commitment Therapy]. Novo Hamburgo, RS: Sinopsys.

Webb, C. A., DeRubeis, R. J., & Barber, J. P. (2010). Therapist adherence/competence and treatment outcome: A meta-analytic review. *Journal of Consulting and Clinical Psychology, 78*(2), 200–211.

Westrup, D. (2014). *Advanced acceptance and commitment therapy: The experienced practitioner's guide to optimizing delivery.* Oakland, CA: New Harbinger Publications, Inc.

Wheeler, M. (2011). Martin Heidegger. In E. N. Zalta (Ed.), *The Stanford encyclopedia of philosophy* (Winter 2012 ed.). Retrieved from https://plato.stanford.edu/entries/heidegger/.

Wickman, F. (2013). Oh, please. When did we start rolling our eyes to express contempt? *Slate.com*. Retrieved August 13, 2017, from http://www.slate.com/articles/life/explainer/2013/01/eye_rolling_why_do_people_roll_their_eyes_when_they_re_annoyed.html.

Wilson, K. G. (2009). *Mindfulness for two: An acceptance and commitment therapy approach to mindfulness in psychotherapy.* Oakland, CA: New Harbinger Publications, Inc.

Yalom, I. (1980). *Existential Psychotherapy.* New York, NY: Basic Books.

Robyn D. Walser, PhD, is codirector of the Bay Area Trauma Recovery Clinic, staff psychologist at the National Center for PTSD Dissemination and Training Division, and assistant clinical professor in the department of psychology at the University of California, Berkeley. As a licensed clinical psychologist, she maintains an international training, consulting, and therapy practice. Walser is developing innovative ways to translate science into practice, with a focus on the dissemination of state-of-the-art knowledge and treatment interventions.

Foreword writer **Steven C. Hayes, PhD**, is Nevada Foundation Professor in the department of psychology at the University of Nevada, Reno. An author of forty-one books and more than 575 scientific articles, he has shown in his research how language and thought lead to human suffering, and has developed acceptance and commitment therapy (ACT)—a powerful therapy method that is useful in a wide variety of areas.

Contributing writer **Manuela O'Connell, Lic**, is an ACT therapist and protégé of Robyn Walser who lives in Buenos Aires.

Contributing writer **Carlton Coulter, DClinPsy**, is a clinical psychologist and specialist in the field of adult mental health with more than fifteen years' experience.

Index

A

about this book, 6–10
acceptance and commitment therapy (ACT): awareness pillar in, 45; body language in, 99–111; challenges faced in, 173–194; developing competency in, 14–21, 201–208; engagement pillar in, 50; growth in practicing, 130–133; head and hands of, 13; heart of, 2, 9, 13; interpersonal process in, 120–121, 171–194; intrapersonal process in, 121–123, 149–169; openness pillar in, 37; overarching context of, 35–36; as personal journey, 8; processes in, 116–124; therapeutic stance in, 28–30, 195–213; three selves in, 61–72; training therapists in, 3–6; unique languaging of, 87–88; wisdom in, 2, 9
Acceptance and Commitment Therapy (Hayes et al.), 7
acceptance-based language, 88, 93
ACT Made Simple, 2nd Ed. (Harris), 7
ACT Practitioners Guide in the Science of Compassion, The (Tirch, Schoendorf, & Silberstein), 7
action, committed, 50, 52–53
Advanced ACT (Westrup), 3, 7
After the Ecstasy, the Laundry (Kornfield), 39
agape, 168
anger, 59–60, 188
Aristotle, 13
attachments: letting go of, 37, 39, 43; practice of exploring, 41–42
audio-recording sessions, 146–147
authentic disclosure, 153–154
avoidance behavior, 178
awareness, 45–49; ACT core processes related to, 45; of death, 73, 75, 76–77, 83–85, 215; flexibility and the practice of, 45–46, 47; listening with, 46–49, 210–212; practice of mindful, 47, 66, 68–69, 199; reflective practice questions on, 47; self-, 49, 57–58, 61, 73, 197–200; of self-as-process, 66–68
awareness pillar in ACT, 45

B

balance in therapy, 27–28
behavior: approaches to understanding, 117–118; recognizing the function of, 35, 124, 126–127; working with suicidal, 190–192
being right, 188–190
beingness, 49, 80
belief, 89–90
big mind dilemma, 180–183
blushing, 109
body language, 66, 99–111; awareness of, 99, 110; clusters of, 102; communication role of, 100; congruence of, 102–104; context of, 100, 101–102; elements of, 108–110; emotion related to, 104–105; questions to consider about, 101; research studies on, 104
bold moves, 205–208
bookending techniques, 25–26
breath, speech related to, 95
burnout, experience of, 165

C

case conceptualization: clinical example of, 136–143; interpersonal process and, 139–141; intrapersonal process and, 141–143; as ongoing process across time, 117–119, 137–139
challenges in therapy, 173–194; being right, 188–190; big giant mind dilemma, 180–183; excessive talking by clients, 178–180; getting stuck in history, 192–193; nothing is good enough, 184–188; silent clients, 183–184; suicidal behavior, 190–192; working harder than your client, 173–180
change: desperation for, 174–175; expressed in action, 82; opening to, 27–31
Chödrön, Pema, 38, 45
choice making, 209
clients: challenges in working with, 173–194; sharing internal experience with, 150, 151, 165–166, 204; stuckness in "fixing," 176–177; style of responding to, 131–132; working harder than, 173–180

clinical examples: of case conceptualization, 136–143; of focus on function, 143–146; of overusing techniques, 22–25
clusters of body language, 102
coherence, 117–118
committed action, 50, 52–53
communication: body language and, 99–111; tone and pace of, 93–99. *See also* language
compassion: ACT in context of, 8; awareness of experience and, 48; boldness related to, 208; core competency of, 203; for human suffering, 172; waning of, 164–166
Compassion Focused Therapy, 7
compassionate immediacy, 83–85
conceptualized self, 61–66, 138
confidence, quality of, 206
congruence of body language, 102–104
conscious being, 45, 111
consciousness, 45, 46, 80, 91, 158. *See also* awareness
contact with the present moment, 45, 183
content: importance of listening to, 129; staying in process vs., 124–128
context: body language, 100, 101–102; definition of, 55; therapeutic, 55, 56. *See also* self-as-context
contextualization, 132–133
contingency-shaped behavior, 16
control: creative hopelessness and, 84, 144, 157; language related to, 87–88, 89, 90; values work vs. agenda of, 52; willingness as alternative to, 24–25; workday example of, 88
core competencies in ACT, 201–208
Coulter, Carlton, 10
countertransference, 58
creative hopelessness, 52, 84, 144–145, 157, 192

D

death: avoidance related to, 182; awareness of, 73, 75, 76–77, 83–85, 215; compassionate immediacy and, 83–85; exercises on exploring, 79, 83, 85; meeting life in the context of, 138–139; self-as-context and, 79–80; spiritual/religious beliefs about, 76; values clarification and, 77–78
defenses, therapist, 61–62, 72
defusion: as avenue to openness, 36, 37; big mind dilemma and, 182
desperation, feelings of, 174–175
discomfort, growth and, 17–19

dispositional mindfulness, 68
Donne, John, 84

E

Einstein, Albert, 218
emotions: body language and, 104–105; critical roles of, 142; identifying your own, 159–160; positive, in session, 167–168; sharing experience of, 150, 151, 165–166; therapist, 58–60, 150, 158–161, 165–166
empathy, loss of, 164–166
engagement, 16, 50–53; ACT core processes related to, 50; committed action and, 50, 52–53; values connected to, 50–52
engagement pillar in ACT, 50
equality, 202
Eugenides, Jeffrey, 159
evoking language, 98
exercises. *See* Reflective Practice exercises
existential issues, 81, 182, 209–210
expectations about therapy, 176
experiential knowledge, 15, 16, 91, 183
eye contact, 108–109
eye rolling, 101–102
Eyes On exercise, 109

F

Five Facet Mindfulness Questionnaire, 68
"fixing" clients, 176–177
flexibility: awareness practice and, 45–46, 47, 73; clinician's compass regarding, 138
fluidity, quality of, 218
Frankl, Victor, 50
freedom, 81–82, 209
Frost, Robert, 215
frustration, 59–60, 143, 160
function of behavior: content vs., 124–128; example of focus on, 143–146; importance of understanding, 156; speaking to, 89–91
functional contextualism, 117
Funeral Exercise, 77
fusion, profound, 180–183

G

genuineness, 203–204
gestures, 109
grief, 86
growth: discomfort and, 17–19; in practicing ACT, 130–133; roadblocks to, 21–27

H

hands of ACT, 13
Hayes, Steven C., x, 26, 84
head of ACT, 13
heart of ACT, 2, 3, 9, 13
hierarchical framing, 139
holding the tension, 204–205
homework assignments, 175
hopelessness, creative, 52, 84, 144–145, 157, 192
hopes and dreams: getting stuck in, 192–193; giving up, 37–38, 39, 41, 43, 44
humility, therapist, 51

I

"I can't" response, 137–139
I/here/now awareness, 45, 72, 73
I/there/then awareness, 73
immediacy, compassionate, 83–85
intention, therapeutic, 28–30
internal experience: language reflecting acceptance of, 89; sharing with clients, 150, 151, 165–166, 204
interpersonal process, 120–121, 171–194; case conceptualization in, 139–141; challenges in therapy and, 173–194; focus on function in, 145–146; suffering as obstacle in, 171–172. *See also* therapeutic relationship
interrupting clients, 179, 180
intimidation, feelings of, 162–164
intrapersonal process, 121–123, 149–169; case conceptualization and, 141–143; focus on function in, 145–146; personal experience and, 161–167; positive emotion and, 167–168; therapist self-disclosure and, 150–155, 167; willingness related to, 155–161
irreverence, use of, 90, 180

J

Jacques, Brian, 86
jargon, avoiding, 96–97

K

Kabat-Zinn, Jon, 199
Keller, Helen, 149
King, Martin Luther, Jr., 1
knowledge: pursuit of self-, 57–60; verbal vs. experiential, 15
Kornfield, Jack, 39

L

La Bruyère, Jean de, 217
language: acceptance-based, 88, 93; awareness of words and, 81; consciousness and, 91–92; control-based, 87–88, 89, 90; emotion and, 159; jargon in, 96–97; mindful use of, 97–98; pace of speech and, 93, 94–98; speaking to function, 89–91; tone of voice and, 93–94; unique to ACT, 87, 88. *See also* body language
Lao Tzu, 55, 195, 216
learning ACT: challenges in, 15; therapist training for, 3–6
Learning ACT, 2nd Ed. (Luoma et al.), 7, 117, 201
letting go, 27, 37, 39, 43
listening with awareness, 46–49, 210–212
loneliness, clinical example of, 143–146
love, expression of, 168

M

Man's Search for Meaning (Frankl), 50
Mastering the Clinical Conversation (Villatte, Villatte, & Hayes), 7
meaning: personal creation of, 50; staying connected to, 52, 85
meditation experience, 68–69
Mehrabian, Albert, 100
mind: awareness of, 158; big mind dilemma, 180–183; excessive use of, 27; fusion with, 90; observing, 92; suffering related to, 27; of therapist in session, 155–158
Mindful Couple, The (Walser & Westrup), 43
mindfulness: practice of, 47, 66, 68–69, 199; speaking with, 97–98; in therapy journey, 183
Mindfulness for Two (Wilson), 7
mindfulness retreat story, 37
Miracle of Mindfulness, The (Nhat Hanh), 91
monkey mind, 91, 158
motion, being in, 35–36

N

Nhat Hanh, Thich, 91, 158
no arrival, 15, 32, 115, 130, 215
nonattachment, practice of, 37–38, 43, 44
nonverbal cues, 100, 101, 102–103. *See also* body language
nonverbal sensitivity, 104
"nothing is good enough" behavior, 184–188

noticing: of personal attachments, 42, 92; "well noticed" technique and, 22–24

O

Obama, Barack, 135
observing mind, 92
obstacles to therapy, 171, 172–173
O'Connell, Manuela, 10
ongoing therapy process. *See* overarching and ongoing therapy process
openness, 36–45; ACT core processes related to, 37; letting go as avenue to, 37, 39, 43; personal rise and fall of, 39–41; as way of being, 42–45
openness pillar in ACT, 37
organizing principles, 118, 119
overarching and ongoing therapy process, 116–120; case conceptualization in, 117–119, 137–139; focus on function in, 144–145; therapeutic relationship in, 119–120
overwhelm, feelings of, 166–167

P

pace of speech, 93, 94–98
pain, transformation of, 217
Pardo, Jennifer, 93
patience, 184
pauses in therapy, 156
peak-end bias, 195–196
personal willingness. *See* willingness
perspective taking, 69, 199
philosophical approaches, 117–118
pillars of ACT: awareness pillar, 45; engagement pillar, 50; openness pillar, 37
politeness rule, 125
positive emotions, 167–168
power differential, 62
presence, therapeutic, 2, 196, 212
present-moment contact, 45, 183
problem solving, 178, 190, 192
processes in ACT, 116–124; interpersonal process, 120–121, 171–194; intrapersonal process, 121–123, 149–169; overarching, ongoing process, 116–120
process-oriented therapy, 136
purpose, living with, 85

R

recording sessions, 146–147
reflection, 15–16, 19

Reflective Practice exercises, 10; on body language, 111; on boldness in therapy, 208; on challenges in therapy, 193–194; on ego exploration, 53; on emotional reactions, 60; on exploring your death, 79, 83, 85; on letting go of conceptualizations, 80; on mindfulness practice, 47, 69; on noticing attachments, 41–42, 92; on pace and tone of speech, 99; on positive emotions in session, 168; on recognizing organizing principles, 119; on recording and reviewing sessions, 146–147; on reviewing your clinical work, 31, 123–124; on self-as-context, 72; on self-awareness, 199, 200–201; on self-disclosure in therapy, 155, 167; on stories interfering with therapy, 62; on talking less in sessions, 158; on therapy self-assessment, 21, 133
religion and spirituality, 76
responding to clients: confronting your style of, 131–132; listening and, 210–212
responsibility, 81–83, 209
right vs. wrong dilemma, 188–190
Rogers, Carl, 49, 115
rule-governed behavior, 16

S

sadness, 86, 143
Santayana, George, 77, 215
self: conceptualized, 61–66, 138; perspective taking sense of, 69
self-as-context, 69–72; awareness and, 45; death and, 79–80; explaining vs. experiencing, 157–158
self-as-process, 65–69
self-awareness, 49, 57–58, 61, 73, 197–200. *See also* awareness
self-care, 165
self-compassion, 92
self-content, 64–65
self-disclosure, 142, 150–155; authenticity in, 153–154; purposes served by, 151–152, 153; reflective practices on, 155, 167
self-knowledge, 57–60, 122
self-reflective perspective, 198
sessions of therapy. *See* therapy sessions
sharing internal experience, 150, 151, 165–166, 204. *See also* self-disclosure
silence: problematic, 183–184; promoting periods of, 179–180; usefulness of, 155–156
skin tone changes, 109

slowing down, 95, 97, 131, 178
small attachments: letting go of, 39, 40; practice of exploring, 41–42
soft tone of voice, 93
space making, 36
spatial presence and distance, 109–110
specialness, myth of, 83
speech: mindful use of, 97–98; pace of, 93, 94–98; pauses in, 156
spirituality and religion, 76
stance of therapist. *See* therapeutic stance
state mindfulness, 68
stories: client, 129; therapist, 62–64
suffering: personal acknowledgment of, 13; related to mind and language, 27, 34; therapist response to, 171–172
suicidal behavior, 190–192

T

talking, excessive, 178–180
techniques: balancing the use of, 27–28; bookending or contextualizing, 25–26; examples of overusing, 22–25; integrating and embedding, 31
tenacity, therapist, 183
tensions, holding, 204–205
therapeutic presence, 2, 196, 212
therapeutic relationship: case conceptualization in, 139–141; challenges in, 173–194; context of, 55, 56; continual evolution of, 119–120; focus on function in, 145–146; illustrated model of, 202; interpersonal process in, 120–121; maintaining balance in, 27–28; mindful awareness in, 68; stance in, 195–213; suffering as obstacle in, 172–173
therapeutic stance, 28–30, 195–213; deepening or expanding, 212; definition and explanation of, 195, 196; developing core competencies for, 201–208; existential angst and, 209–210; listening and responding in, 210–212; self-awareness related to, 197–200
therapists: challenges for, 173–194; growth in ACT practice, 130–133; in-session emotions of, 58–60, 150, 158–161; in-session mind of, 155–158; intimidation felt by, 162–164; overwhelm felt by, 166–167; personal willingness of, 155–161; response style of, 131–132; self-disclosure by, 142, 150–155, 167; stories interfering with work of, 62–64;

training in ACT, 3–6; waning of compassion in, 164–166
therapy sessions: emotions of the therapist in, 58–60, 150, 158–161; mind of the therapist in, 155–158; positive emotions in, 167–168; recording and reviewing, 146–147
Thoreau, Henry David, 87
three selves in ACT, 61–72; conceptualized self, 61–65; self-as-context, 69–72; self-as-process, 65–69
timing of self-disclosure, 152
tone of voice, 93–94
training therapists in ACT, 3–6; current state of, 4–5; future of, 5–6
trait mindfulness, 68
Treasure, Julian, 48

U

unpredictability, patterns of, 138
urgency, compassionate, 83–85

V

values: death and clarification of, 77–78; living with purpose and, 85; personal engagement and, 50–52
verbal knowledge, 13, 15, 92
victimhood, identification with, 185–188
video-recording sessions, 146–147
voice, tone of, 93–94
vulnerability, 202–203

W

walking together, 219
"well noticed" technique, 22–24
Westrup, Darrah, 3, 43
Whitford, Bradley, 33
willingness: as alternative to control, 24–25; as avenue to openness, 36, 37; and emotion of therapist in session, 158–161; and mind of therapist in session, 155–158
wisdom in ACT, 2, 3, 9
words: listening beneath the, 46–49; recognizing arbitrariness of, 81. *See also* language

Y

Yalom, Irvin, 84

MORE BOOKS *from* NEW HARBINGER PUBLICATIONS

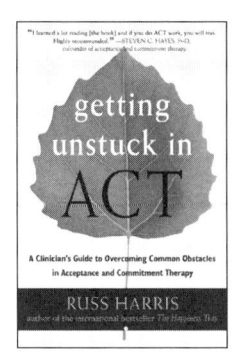

GETTING UNSTUCK IN ACT
A Clinician's Guide to Overcoming Common Obstacles in Acceptance & Commitment Therapy
978-1608828050 / US $29.95

COGNITIVE DEFUSION IN PRACTICE
A Clinician's Guide to Assessing, Observing & Supporting Change in Your Client
978-1608829804 / US $39.95
CONTEXT PRESS
An Imprint of New Harbinger Publications

ACT QUESTIONS & ANSWERS
A Practitioner's Guide to 150 Common Sticking Points in Acceptance & Commitment Therapy
978-1684030361 / US $29.95
CONTEXT PRESS
An Imprint of New Harbinger Publications

LEARNING ACT FOR GROUP TREATMENT
An Acceptance & Commitment Therapy Skills Training Manual for Therapists
978-1608823994 / US $59.95
CONTEXT PRESS
An Imprint of New Harbinger Publications

THE ESSENTIAL GUIDE TO THE ACT MATRIX
A Step-by-Step Approach to Using the ACT Matrix Model in Clinical Practice
978-1626253605 / US $59.95
CONTEXT PRESS
An Imprint of New Harbinger Publications

THE BIG BOOK OF ACT METAPHORS
A Practitioner's Guide to Experiential Exercises & Metaphors in Acceptance & Commitment Therapy
978-1608825295 / US $59.95

newharbingerpublications
1-800-748-6273 / newharbinger.com

Follow Us

(VISA, MC, AMEX / prices subject to change without notice)

Quick Tips *for* Therapists
Fast and free solutions to common client situations mental health professionals encounter every day

Written by leading clinicians, Quick Tips for Therapists are short e-mails, sent twice a month, to help enhance your client sessions. Visit newharbinger.com/quicktips to sign up today!

Sign up for our Book Alerts at **newharbinger.com/bookalerts**

PRAXIS
MANY VOICES | ONE WORK
A subsidiary of New Harbinger Publications, Inc.

Enhance your practice with live ACT workshops

Praxis Continuing Education and Training—a subsidiary of New Harbinger Publications—is the premier provider of evidence-based continuing education for mental health professionals. Praxis specializes in ongoing **acceptance and commitment therapy (ACT)** training—taught by leading ACT experts. Praxis workshops are designed to help professionals learn and effectively implement ACT in session with clients.

- **ACT BootCamp®: Introduction to Implementation**
  For professionals with no prior experience with ACT, as well as those who want to refresh their knowledge

- **ACT 1: Introduction to ACT**
  For professionals with no prior experience with ACT

- **ACT 2: Clinical Skills-Building Intensive**
  For professionals who practice ACT, but want more hands-on experience

- **ACT 3: Mastering ACT**
  For professionals actively using ACT who want to apply it to their most complex cases

Receive 10% off Any Training!
Visit praxiscet.com and use code **EDU10** at checkout to receive 10% off.

Check out our workshops and register now at praxiscet.com

Can't make a workshop? Check out our online and on-demand courses at **praxiscet.com**

CONCEPTUAL | EXPERIENTIAL | PRACTICAL

Register your **new harbinger** titles for additional benefits!

When you register your **new harbinger** title—purchased in any format, from any source—you get access to benefits like the following:

- Downloadable accessories like printable worksheets and extra content
- Instructional videos and audio files
- Information about updates, corrections, and new editions

Not every title has accessories, but we're adding new material all the time.

Access free accessories in 3 easy steps:

1. Sign in at NewHarbinger.com (or **register** to create an account).

2. Click on **register a book**. Search for your title and click the **register** button when it appears.

3. Click on the **book cover or title** to go to its details page. Click on **accessories** to view and access files.

That's all there is to it!

If you need help, visit:

NewHarbinger.com/accessories

new harbinger
CELEBRATING
40 YEARS